THE PEATLANDS
OF BRITAIN AND IRELAND
A TRAVELLER'S GUIDE

CLIFTON BAIN

With drawings by Darren Rees
Foreword by Tony Juniper

SANDSTONE PRESS
HIGHLAND | SCOTLAND

First published in Great Britain and Ireland in 2021.
Sandstone Press Ltd
PO Box 41
Muir of Ord
IV6 7YX

www.sandstonepress.com

All rights reserved.
No part of this publication may be reproduced, stored or transmitted in any form without the express written permission of the publisher.

© Clifton Bain 2021
© Foreword 2021 Tony Juniper
© Drawings Darren Rees 2021
© Images as acknowledgements
© All maps RSPB 2021. Contain Ordnance Survey data © Crown copyright and database right 2019 © OpenStreetMap contributors, UKCEH Land Cover 2019, Corine Land Cover 2012 © European Environment Agency.
Editor: Alison Lang

The moral right of Clifton Bain to be recognised as the author of this work has been asserted in accordance with the Copyright, Designs and Patents Act 1988.

The publisher acknowledges support from IUCN UK Peatland Programme, and the RSPB towards the publication of this volume.

ISBN: 978-1-912240-24-1

Cover and book design by Heather Macpherson at Raspberry Creative Type, Edinburgh.
Printed and bound by Elma Printing & Finishing

FOREWORD

A Modern Perspective on Peatlands

Britain and Ireland's peatlands comprise some of our most precious and evocative landscapes. From the vast expanses of blanket bog that smother the broad, undulating vistas of west coast Ireland and the north of Scotland to the valley mires of the southern heaths, and from the raised bogs of the north-west of England to the Fens of East Anglia, these wonderful natural systems are diverse in character.

My own forays as a naturalist and conservationist have taken me to many of these places, very often as part of some effort to save them from damage or destruction. Like many others, my motivation came from a growing appreciation of the wonderful wildlife that depends on such ecosystems, and also an appreciation of the unique circumstances that cause them to exist in the first place.

Peat is formed in wetlands, where water causes the decomposition of dead vegetation to be exceeded by its accumulation, over time leading to ever deeper deposits of plant material. This and other unique aspects of peatlands have long fascinated ecologists, although today it is increasingly clear how peatlands are not only of great interest, but also of great importance. Although for too long they have been regarded as desolate wastelands, we know now that peatlands are in fact hugely valuable.

Despite our growing appreciation of their significance, however, many peatlands remain under pressure and are in a process of progressive degradation. As they reveal the effects of deteriorating health, so damaged peatlands fulfil fewer of their previous functions, in turn leading to diminished value for both society and wildlife.

Rising carbon dioxide concentration in the air, the inundation of fields and homes after heavy rain, poor water quality and disappearing wildlife are all among the consequences, in turn creating costs and reduced value for people.

The good news is that not only can healthy peatlands be conserved, but damaged ones can be restored, in the process realising many benefits. In the pages that follow, readers can learn more about not only the nature and character of Britain and Ireland's rich peatlands, but also the ways in which it is possible to place peatlands on the road to recovery, in the process reversing the effects of centuries of ignorance as to their true value.

Clifton Bain tells the story of these wonderful ecosystems not only with passion and clarity, but also with the kind of unparalleled insight that can only be drawn from more than three decades of study and advocacy. Vivid first-hand accounts bring the peatlands to life, and should the reader be inspired to get their boots on and get outside to see for themselves, then detailed maps will take you to where you need to go.

While the story of our peatlands is at times one of decline and damage born out of short-term expediency and a failure to see the bigger picture, this book gives life to a more modern perspective, wherein healthy peatlands can support a sustainable future, not only for wildlife but people too. More important still is the message that we have the means to restore peatlands, if only we choose to adopt them. That choice is one for society as a whole, which in turn will depend upon our collective awareness, and that is why this book is so important.

Dr Tony Juniper CBE, environmentalist and writer

ACKNOWLEDGEMENTS

The Peatlands of Britain and Ireland is intended as a celebration of the remarkable turnaround in the fate of our peatlands as society begins to appreciate their true worth. Throughout my career, I have met many experts and individuals committed to peatland conservation who have helped bring about this transformation and who have guided me on my continued journey of learning and enjoyment.

Working with the IUCN UK Peatland Programme has brought my peatland understanding to a level that has enabled this book to be produced. All those involved in that organisation deserve special mention for their commitment to peatland conservation. Integral partners include the Wildlife Trusts, specifically Yorkshire and Scottish Wildlife Trusts, and the RSPB, which between them have hosted the programme over the last ten years.

The programme has been bolstered by dedicated members of staff, currently Emma Goodyer, Sarah Proctor, Blue Kirkhope and Renée Kerkvliet-Hermans, and previously Joanna Richards, Jillian Hoy, Lyndon Marquis, Rea Cris and Mary Church, with chairs Rob Stoneman and Jonathan Hughes representing the Wildlife Trusts, and Stuart Brooks, formerly of the John Muir Trust and now with National Trust for Scotland. Long-standing steering group members include Pat Thompson and Olly Watts from the RSPB, Paul Vaight, formerly a Peter De Haan Charitable Trust trustee, Moors for the Future Partnership manager Chris Dean, Paul Leadbitter of the North Pennines Area of Outstanding Natural Beauty Partnership, Andrew Coupar and Andrew McBride of NatureScot (previously called Scottish Natural Heritage), Ian Crosher and Iain Diack of Natural England, Peter Jones of Natural Resources Wales, Tim Thom of Yorkshire Wildlife Trust, Bruce Wilson of Scottish Wildlife Trust, Ben McCarthy of the National Trust, David Smith of South West Water, Richard Lindsay of the University of East London (my long-standing font of wisdom) and Professor Mark Reed of Scotland's Rural College.

Their efforts are supported by a wider steering group that includes the UK Department for Environment, Food and Rural Affairs (DEFRA), Natural England, the Welsh Government, Natural Resources Wales, the Scottish Government, NatureScot and the Northern Ireland Department of Agriculture, Environment and Rural Affairs (DAERA). Representatives from many of these organisations have been present from the start and their collaboration has been essential in driving peatland conservation forward.

Special thanks to Des Thompson at NatureScot for once again providing wise advice just when needed and who firmly believed this book had to be written.

I am grateful to Ellen Wilson and Paul Britten for providing the maps and to Carl Mitchell and Mick Green as companions on the site visits as well as being eagle-eyed fact checkers.

My thanks to all those in my team and the steering group who gave advice and assistance in checking sections of text and to the others who kindly gave their help: Morag Angus, Robbie Carnegie, Brian Eversham, Catherine Farrell, Ben Gearey, Benjamin Inglis-Grant, Paul Harvey, Peter Jones, Helen Lawless, Nuala Madigan, Isabella Mulhall, Norrie Russell, Steve Sankey, Fiona Walker and Sue Whyte.

I am indebted to the many people who offered advice and guidance while planning and undertaking my site visits, including Matt Buckler, Peter Coldwell, Rebecca Dobson, Catherine O'Connell, Dave O'Hara, Mark McCorry, Chris Uys, Tristram Whyte and Arfon Williams.

While I managed to complete all my site visits before a global pandemic forced nationwide lockdowns, I was unable to complete the photography for all sites and I appreciate the help of all those who kindly provided images: Penny Anderson, Emma Austin, Dave Blackledge, Laurie Campbell, Robbie Carnegie, Iain Diack, Brian Eversham, Christine Hall, Lorne Gill, Benjamin Inglis-Grant, Eden Jackson, Peter Jones, Blue Kirkhope, Alan Leitch, Lyndon Marquis, the Moors for the Future Partnership, the National Museum of Ireland, Norrie Russell, Mike Perks, Shropshire Wildlife Trust and the Yorkshire Peat Partnership.

My special thanks to Robert Davidson, Alison Lang and Nicola Torch at Sandstone Press and Heather Macpherson at Raspberry Creative Type for seeing through this third book in the series, and to Darren Rees for once more providing such excellent artwork.

To my wife Karin, thank you as always for your patience and support and thanks also to my daughters Ellie and Kirsten who seem to have wriggled out of a trip to the bogs, so far.

Dubh Lochan Trail, Forsinard Flows

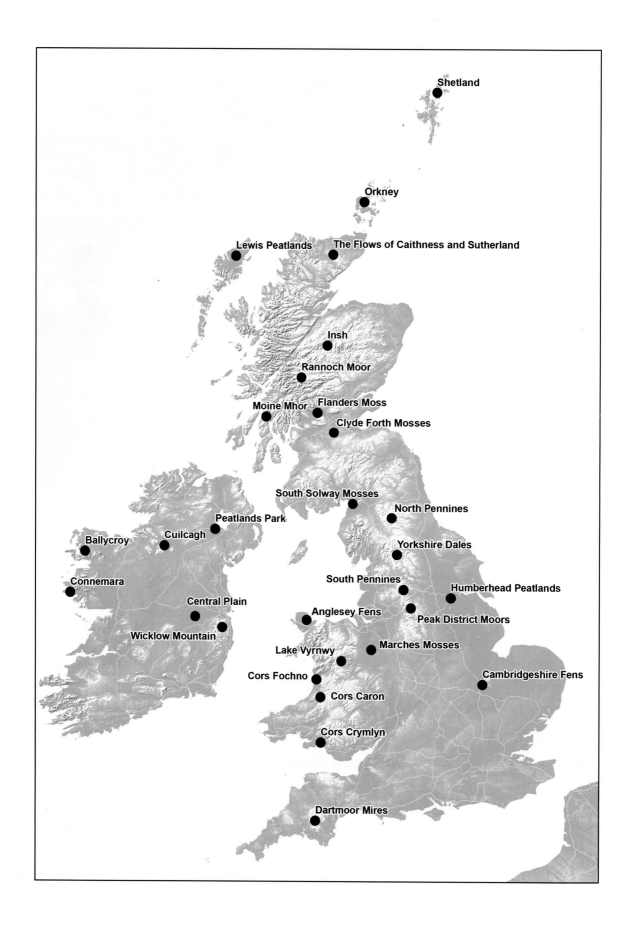

CONTENTS

Foreword by Tony Juniper	3
Acknowledgements	5
Contents	9
Introduction	11
The Peatlands of Britain and Ireland	20
Peatland	21
Bogs	24
Fens	27
Wildlife	27
Bog wildlife	27
Fen wildlife	35
Peatland archaeology	39
Bodies in the bog	41
Trackways	43
Preserved landscapes	44
Lost archives	45
From lost peat to lost words	45
Peatland Use	49
Bog iron	49
Peat as fuel	49
Commercial peat extraction	52
Cultivated peatlands	54
Built development	58
Peatlands' true benefits	59
A new era for peatlands	63
Visiting peatlands	69
Using this guide	73
Key to maps	73
SCOTLAND	74
North-West Highlands and Northern Isles	76
Shetland	79
Orkney	85
The Flows of Caithness and Sutherland	91
Outer Hebrides	96
Lewis Peatlands	99
Grampian Mountains	104
Insh Marshes	107
Rannoch Moor	111
Moine Mhòr	115
Central Lowlands	118
Flanders Moss	121
Clyde-Forth Mosses	127
ENGLAND	132
Northern England	134
Peak District Moors	137
Yorkshire Dales & South Pennines	143
North Pennines	149
Midlands	152
Humberhead Peatlands	155
Marches Mosses	161
North-West England	166
South Solway Mosses	169
Eastern England	174
Cambridgeshire Fens	177
South England	182
Dartmoor Mires	185
WALES	190
North Wales	192
Anglesey Fens	195
Lake Vyrnwy	201
Mid Wales	204
Cors Fochno	207
Cors Caron	211
South Wales	214
Cors Crymlyn	217
IRELAND	220
Northern Ireland	222
Peatlands Park	225
Cuilcagh Mountain	229
Republic of Ireland	232
Ballycroy	235
Connemara	239
Central Plain	243
Wicklow Mountains	249
Safety and access	252
Selected bibliography	253
Useful websites	254
Photo credits	255

INTRODUCTION

People are often unaware that peatlands can be found close to most of our urban centres, providing a relaxing, uncluttered, airy escape. They also form some of our most dramatic remote landscapes, where not a single built structure can be seen from horizon to horizon. The mental and physical health benefits of walking or cycling among such wide-open spaces are but some of the assets of this fascinating environment. Although I have studied peatlands throughout my career, I am still enthralled and energised by my visits. The tremendous achievements made in recent years towards securing the future of our peatlands allow us all the opportunity to enjoy a day out on the bog. As with my other two books, *The Ancient Pinewoods of Scotland* and *The Rainforests of Britain and Ireland*, this latest guide encourages readers to get out and experience some of our greatest natural wonders while better understanding their importance.

This book also marks a celebration, recognising the support of the numerous partners across government, conservation bodies, scientists and private land managers who have helped us firmly enter the new era for peatlands.

A keen hillwalker since my teens, my acquaintance with peatlands was initially a casual one. Boggy ground was something to be carefully avoided when on the moors. I would admire the spongy, colourful mosses and intriguing carnivorous plants but otherwise gave these sodden places little thought. However, peatlands would later become a frequent staging post in the development of my career, more by chance than intention.

My first job after university was with the RSPB in North Wales, investigating the decline in breeding birds on peatlands in the Denbigh Moors. Moving back to Scotland in the late 1980s, I was a foot soldier in the environmental campaign to stop commercial forestry planting from destroying the Flow Country peatlands of Caithness and Sutherland. The issue attracted considerable media attention, partly through the involvement of celebrities who had financed the new forests as a tax benefit, unaware of the environmental harm being caused.

One of my tasks at the RSPB was to catalogue the results of breeding bird surveys carried out across the Flows. It was hard to comprehend some of the figures, with thousands of pairs of breeding wading birds being recorded. Sitting at my desk in Edinburgh, I couldn't wait for my first chance to experience such a concentration of wildlife.

At that time, few people had heard of the Flows. This vast, remote peatland in the far north of the Scottish mainland was out of sight and out of mind even to those living in

Owenduff Bog, County Mayo

nearby towns. A first visit to the region cannot fail to be an emotional one. The train from Inverness to Thurso goes through the heart of the Flows. The lonely carriages are a mere speck in the landscape and pass mile after mile of open peatland, uninterrupted from horizon to horizon apart from the intrusion of those ill-conceived conifer plantations. At any time of year, the huge expanse of mosses patterned by pools of water is a spectacular treat, but it is the spring and summer months that offer the greatest reward, when the peatlands come alive with the calls of birds. Now officially recognised as the best example of an Atlantic blanket bog in the world and supporting some of the largest populations of peatland bird species in Europe, it is a travesty that this area was targeted for forestry.

The Flow Country controversy brought out some of the most dedicated and effective conservation efforts I have ever seen. Individuals persisted in championing this important cause despite personal and institutional abuse. I learned some important lessons at that time. Economics are a key force affecting our environment. Understanding the financial implications behind land use decisions is essential in resolving conflicts.

An equally important lesson was the appreciation that local people are vital in bringing about long-term solutions to environmental problems. Residents of villages and towns near the peatlands were initially bemused by attempts to protect such perceived 'wasteland' or were angry at the intervention of outside conservationists in their affairs. Conflict over peatland use across the country saw extreme cases of local protest where effigies of conservationists were strung up on mock gallows. Such acts were usually carried out by individuals fuelled by controversy-hungry media rather than being representative of the community, but they serve to highlight the local tensions at the time.

I also witnessed the power of well-presented, strong scientific evidence in influencing political decision makers. Demonstrating the international significance of the peatlands and their wildlife, as well as proving the economic folly of planting trees on peat, led to success for the Flows campaign. The tax incentives were removed, and protected site status was given to much of the peatland area. Public funding and European Union grants then enabled a long programme of partnership working with environmental bodies, government agencies, landowners and local communities supporting a shared agenda for the future well-being of the peatlands. Challenges remain, with ongoing threats from windfarms and other built development, but there is now a greater

Forsinard Flows, Lookout Tower

recognition that the Flows are something special and among the world's most important natural treasures.

My second phase of peatland conservation was in the 1990s. When working for the RSPB, I visited the lowland peatlands at Thorne and Hatfield Moors in South Yorkshire. A pleasant walk on a sunny summer evening turned into shock when I witnessed the devastation caused by industrial mining of peat to supply growbags for gardeners. I couldn't believe that such a wonderful wildlife site could be stripped of its living layer by huge machines leaving behind an eroded moonscape of drained, bare peat. I learned that our planning laws allowed this destruction. Permission for peat extraction had been granted several decades previously, at a time when the peat was cut by hand. The planning authorities were unable to halt the modern operations without hefty compensation to the peat mining companies.

A major campaign was launched to discourage gardeners from using peat and to halt the destruction of Thorne and Hatfield Moors and many other wonderful peatlands across Britain and Ireland. National conservation charities joined with local naturalists and individual 'defenders of the bog' to challenge the multinational companies who saw profit in selling peat. The celebrity face behind the peatlands campaigning at that time was Professor David Bellamy, who enthused about every living detail of the peatlands and was famous for his television appearances, getting right down among the mosses and bog creatures. In the end, the government conservation agency, Natural England, stepped in to purchase Thorne and Hatfield Moors for the nation, and several other commercially worked peatland sites. Peat mining continues elsewhere, but there is now a government target for the use of peat in gardening and horticulture products to be phased out in the UK by 2030.

My conservation career later moved into areas of policy, such as biodiversity and climate change. I began to appreciate that conserving the natural environment could not be done by arguing the importance of wildlife alone. Additional tools are needed to tackle the overwhelming economic forces that influence change in land use and development pressure. Conservation now highlights that helping nature helps us, whether providing vital services such as jobs, well-being and health benefits or natural resources such as clean drinking water. Economic studies are increasingly being deployed to demonstrate the financial implications of our use and abuse of the natural environment. Quantifying nature in monetary terms does not mean we are selling out or belittling the intrinsic or aesthetic value of wildlife, but is simply ensuring that nature is at least accounted for in the face of harsh economic decision making.

These formative years helped build an approach to environmental issues that recognised the strength of partnership working and the need to understand the pressures behind a problem. Identifying people's motivations for the way in which they manage the countryside allows the development of alternative solutions for delivering their needs while protecting the environment. Sheep farmers whose stock is too heavily concentrated on the peatland and is causing damage may be willing to protect the area if government funding is directed at supporting farmers for having healthy peatlands instead of incentivising ever more sheep. This is a far more constructive solution than simply banning sheep grazing in these areas, which would be costly to monitor and would fan the flames of hostility towards peatlands. As it is, many farmers now welcome assistance in repairing the eroded deep gullies of a damaged bog where livestock could have become trapped.

In 2009 I was offered the chance to be the director of an organisation aimed at conserving peatlands. The year before, a peatland conference organised through the Wildlife Trusts and the Peter De Haan Charitable Trust (PDHCT) had been held in London. The event exposed frustration that the importance of peatlands was not widely appreciated and that efforts to conserve them were fragmented, uncoordinated and often compromised by conflicting policy decisions. The conference concluded that a new partnership should be created to champion peatlands and bring together policy makers, land managers, environmental organisations, and scientists. The Peatland Programme was formed under the umbrella of the UK's national committee of the International Union for the Conservation of Nature – a rather lengthy title, but it provided the credibility of a globally renowned conservation body and an established partnership of government and environmental organisations.

The IUCN UK Peatland Programme (IUCN UK PP) was launched in 2009 with a grant from the PDHCT and at an early stage it brought together organisations with experience in managing large-scale projects to restore peatlands. These restoration initiatives were delivered through broad partnerships of local people, land managers, wildlife charities and public bodies in areas such as the Flows, the Peak District, the North Pennines, Yorkshire and Exmoor. Fantastic work was being carried out in these areas to demonstrate that past damage to peatlands could be tackled cost-effectively. Remedial works were possible even at a large scale, covering thousands of hectares. The experience of the staff who represented these projects was a great asset to the programme. They understood the problems and what needed to be fixed; they had heard all the arguments and overcome them through bringing people together and building consensus towards looking after the peatlands.

Major grants from the PDHCT and more recently the Esmée Fairbairn Foundation have provided core funding for the programme. The IUCN UK PP steering group and the employed staff team provide a constant point of reference and act as a catalyst for all those working on individual projects and initiatives across the UK. Redressing centuries of peatland damage and securing a new era for peatland conservation needs this constancy of effort and it is hoped that an ongoing umbrella body for peatlands can be maintained.

One of the first tasks of the programme was the Commission of Inquiry. This work engaged scientists to assess the state of our peatlands, the impacts of human activities and the

Peat turves and blanket bog, County Mayo

benefits that arise when peatlands are restored and conserved. The inquiry gathered clear evidence of the huge significance of peatlands as a carbon store, for drinking water, for flood management and as homes for important wildlife. Published in 2011, the inquiry report was the largest review of peatland science ever undertaken in the UK, with over three hundred responses to the draft text.

The results were presented in the House of Lords, the Scottish Parliament and the Northern Ireland Assembly. There was a real sense of enthusiasm at these events, with government ministers and members of parliament quickly appreciating the need for action. This led to the UK's four devolved government environment ministers issuing a joint action statement on peatlands, in February 2013, giving a commitment to their conservation. This has been taken further, with each country establishing specific peatland funding schemes to support private landowners in restoring their peatlands.

The UK's first peatland strategy was launched in April 2018, setting a framework for peatland conservation through to 2040. There is still some way to go to ensure all UK countries are working to the same speed. Scotland has been the first to set targets for peatland restoration, with associated funding now embedded in its policies, with England, Wales and Northern Ireland well on the way in establishing their own peatland strategies. Strategies themselves are never the most exciting initiatives but, as ministers come and go, they are vital in providing a long-term commitment, clear targets and benchmarks against which to check progress.

Internationally, the IUCN global body has highlighted the plight of peatlands with a resolution for all countries to ensure peatland conservation strategies are in place. The Food and Agriculture Organisation of the United Nations is helping to promote peatlands and a Global Peatlands Initiative has been launched with the support of the United Nations Environment Programme. The UK is among the world leaders in peatland conservation and has several decades of practical peatland restoration and management experience that it can share with other countries. International networking also allows for reciprocal learning from countries whose peatlands may be different but essentially face similar problems. The IUCN UK PP has produced UK and international 'Demonstrating Success' reports, showcasing different peatland restoration projects. Restoration sites are all included on the 'Peatland Projects' interactive map available on the IUCN UK PP website. The huge practical experience of those involved in repairing peatlands has also been drawn together to provide an online guide to peatland restoration methods.

A key initiative of the IUCN UK PP was the development of the Peatland Code. Recognising that government funding for peatland restoration alone was insufficient to meet the scale of the problem, the Peatland Code was designed to attract private funding. The rationale behind the code is that businesses can pay for the restoration either to support climate change efforts or to secure some benefit such as improved biodiversity, reduced flooding or drinking water improvements. The Peatland Code process provides assurance to potential investors that the projects can deliver the claimed climate change or other benefits. Increasingly, businesses are appreciating that their financial bottom line is affected by the state of our environment and that supporting nature can save them money in the long term. It is hoped that the benefits of repairing and looking after a bog can become as popular and widely appreciated as planting trees.

As a partnership body, the IUCN UK PP operates across dozens of projects and engages individuals from many different sectors. A series of annual conferences provide platforms for sharing experience and allowing an all too rare opportunity for scientists, policy makers, land managers and peatland practitioners to get together and see peatland conservation work at first hand. Each year, the conference has moved around the UK, starting in Durham, and has been well attended and enjoyed. As one renowned peatland expert, Richard Lindsay, noted, 'after years of working with a little regarded topic on the fringes, it's amazing to see so many people gathered to talk about peatlands.'

Through this broad partnership, numerous organisations are now championing peatlands, including land-managing interests such as the sporting estates, water companies and even the Forestry Commission. The spread of involvement stretches from the north of Scotland to South-West England, Wales and Ireland, and also to the far reaches of the Falkland Islands, the most peat-dominated of the UK Overseas Territories.

Decision makers and wider society are beginning to understand the costly consequences of damaging our peatlands and the true benefits of keeping them healthy. There is much more education and engagement to be done to raise awareness of the existence and importance of peatlands in the public view. At a time of great economic upheaval and austerity, getting peatlands on the public radar is challenging but a key focus of the programme.

If peatlands are to continue to be valued and recognised as important by decision makers and land managers, we all need to show our support and appreciation. Only through people experiencing and understanding the true importance of our peatlands can the investment of taxpayers' money be justified to restore and maintain them. Research into remote rural communities' views of peatlands shows that, where people become involved in looking after and managing a peatland, they develop a greater understanding and appreciation of its qualities. Many rural communities have a sense of stewardship and are happy to support the protection of peatlands once they understand the issues, even though this may mean a change from some older ways of managing them.

I am a great believer in getting people from all walks of life to visit peatlands as the best way to create a lasting impression. It doesn't always go to plan, however. Falling into a bog is a rite of passage for even the most experienced peatland expert, and a great social leveller. Several times I've over-enthusiastically stepped off a boardwalk and landed up to my waist in peat, along with my bosses, directors of peat mining companies, television presenters and civil servants. One thing for sure is that such intimate experience of a peatland will long be remembered.

Most peatland visits are safe, pleasurable and dry, and the more that people visit them and find out what they are truly like the better. Thanks to innovative and inspiring peatland conservation and restoration initiatives there are many peatlands to visit across Britain and Ireland with excellent all-ability access. The sites detailed in this book represent just a sample, to give the reader a flavour of the huge diversity of peatland experiences on offer. The Peatland Projects map available on the IUCN UK PP website displays a wider range of peatland sites, including Eyes on the Bog locations at which peatland condition is being monitored as part of a long-term citizen science initiative.

Teal (*Annas crecca*) on Dubh Lochan, Forsinard Flows

The Peatlands of Britain and Ireland

Often described as the Cinderella habitat, peatlands have long been considered worthless, even malevolent, or simply a resource to exploit. Yet they are immensely important to our well-being and can display great beauty. These enchanting, saturated, watery landscapes can at first appear rather muted, with wide vistas displaying only pastel shades of browns and greens, but closer inspection reveals a wealth of colour and pattern, rich in the spectacle and sounds of unusual wildlife. This natural state of a peatland contrasts with the blackened, bare, eroding expanses that have been damaged in an often-failed attempt to make peatlands profitable. Our shocking treatment of this wonderful part of our natural environment not only threatens wildlife but has left a legacy of degradation that now imposes great cost on society as we lose the natural benefit of peatlands.

Peatlands are characterised by waterlogged conditions that restrict decay and allow dead plant material to build up over time as peat. Blanketing our mountain tops and engulfing low-lying land, peat is one of our most abundant soils, not surprisingly in such a persistently wet country. Peatlands also pervade our culture, from the drama of *Wuthering Heights* and Sherlock Holmes's canine mystery on Dartmoor to the aromatic basis for whisky and the modern use of peat in gardens. Widely known, but now practised by few, is the craft of turf cutting to provide fuel in remote rural areas.

Harestail cotton grass, Hatfield Moors

Going back over millennia, the association of people with peatlands has been uniquely captured by their excellent preserving qualities that have allowed us to come face to face with the actual bodies of our ancestors as well as incredible cultural artefacts. One of my earliest associations with peat was from my father's bookcase in the form of a small paperback book, *The Bog People* by P. V. Glob, with its captivating cover of the perfectly preserved Tollund Man who had lived over 2,000 years ago. The serene, calm face belied the fact that this individual had been hanged and placed in the bog as part of an Iron Age ritual.

The peatland story is one of contradictions. Often disregarded as wasteland, peatlands are immensely valuable. Visions of dangerous, boggy swamps with their derogatory associations contrast with the reality of colourful carpets of mosses bejewelled by clear pools of water and hemmed with delicate, white cotton-grass heads.

Over the centuries, the draining and clearing of our peatlands has been one of the most extensive acts of environmental destruction ever imposed on this country. Worldwide, the situation is just as desperate in many other hotspots of human population, where extensive peatlands in Europe, America and South East Asia have been drained and exploited. Global news coverage has shown the human suffering resulting from huge fires on drained peatlands in Indonesia and Russia extending over thousands of kilometres. The economic damage from these fires was estimated at several billion US dollars.

We are now beginning to understand the full costly consequences to society of our peatland legacy. Global leaders herald their importance and action is being taken to conserve them. Huge projects are underway to repair damaged peatlands and reinstate their watery conditions, to allow wildlife to thrive and help secure the benefits we can all derive from them.

With awareness of their international conservation importance there has been considerable investment by governments and environmental charities to provide protected sites with excellent visitor facilities, offering the opportunity to get into the heart of these wildlife treasuries.

Peatland

The term 'wetland' is used to describe areas inundated or saturated by water and where vegetation is specially adapted to thrive in the waterlogged soils. Peatlands are a type of wetland where conditions limit the breakdown and decay of dead plant material, which then accumulates to form peat. In a natural, wet state the peat continues to build up year after year and can become several metres deep. Remarkably, peat under these conditions contains fewer solids than milk, yet we can walk on it, albeit tentatively, due to the structure of the preserved plant material that forms a supporting lattice. This feature of being neither fully solid earth nor liquid water, an in-between place, may explain the ancient reverence for peatlands as places for pre-Christian rituals and religious ceremonies.

For many people, peatlands are associated with the dark stuff: the loose, fluffy compost in a growbag or the aroma of burning peat turves. But this is just the peat soil, the mainly dead part of the system. The key feature is the living layer of plants that thrive in wet conditions, maintaining the peat deposit and continuously adding new material. Peat is generally formed at a rate of one millimetre a year, which means that material laid down around the time of the Viking occupation lies only a couple of spade depths from the peatland surface. Where a peatland has its layer of natural vegetation and is laying down peat, the area is called a mire. If the living layer has been destroyed, for example through the peatland being drained, the deep peat deposits are still called peatlands but are no longer peat-forming mires.

The accumulation of peat occurs where the production of plant material exceeds decay from the action of bacteria and fungi. Water saturation leads to anaerobic (without oxygen) and cooler conditions that inhibit the decay. Water balance is critical, with levels having to remain stable just below the surface of the plant layer for the system to form peat. Too shallow and oxygen is available for the decomposing organisms; too deep and the plants become submerged and can't grow. In colder and wetter climates of the world, the predominant peat-forming plants are the mosses and the sedges as well as 'cushion plants' in the Southern Hemisphere. In hotter, subtropical and tropical parts of the world, peat is composed largely of the roots of grasses, rushes or rush-like species and trees.

As plants grow they remove carbon from the atmosphere through photosynthesis and this is stored in the structure of the plant. The preserved plant material in peat retains that plant carbon but, when exposed to oxygen through damage to the peatland, bacterial decay breaks down the peat and releases carbon dioxide into the atmosphere. It is this release of carbon that is of such concern in the fight against climate change. UK peatlands alone store over three billion tonnes of carbon. If only a fifth of that were released into the atmosphere it would equate to the country's entire annual carbon emissions from all our activities.

Peatland distribution is mainly determined by climate and topography. They are especially abundant in the cold regions of northern Europe and in the wet oceanic and humid tropical countries. Where the climate is less than ideal, peat can still form in localised areas where landscape features allow water to collect, such as on flat, level land. Incongruous as it seems, even Australia and Uganda have their peatlands.

Globally, peatlands cover only three per cent of the land area but are known to occur in ninety per cent of the countries of the world. It is said that the other ten per cent of countries just haven't found their peatlands yet. As recently as 2012, a deep peatland larger in extent than England was discovered in the Congo Basin.

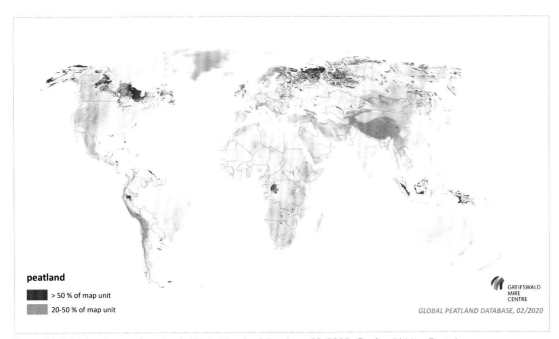

Figure 1 World distribution of peatlands (Global Peatland Database 02/2020, Greifswald Mire Centre).

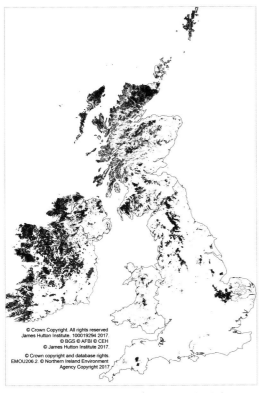

Figure 2 Extent of peatland in Britain and Ireland (from Artz, R. et al 2019. The State of UK Peatlands: an update, IUCN UK Peatland Programme, Edinburgh; and Connolly, J and Holden N.M. 2009. Mapping peat soils in Ireland: updating the derived Irish peat map. Irish Geography 42. 343 - 352)

Britain and Ireland with their northern latitude and cool, wet climate are among the world's most peat-dominated countries. Just over a quarter of Ireland is covered by peatland. In the UK, ten per cent of the land area is currently peatland mire, with the majority in Scotland, supporting some of the deepest peat deposits in Europe – up to ten metres deep in parts. In the UK Overseas Territories, the Falkland Islands are over ninety-four per cent peatland, making them the most peat-dominated of any country in the world. At one time, Britain was almost a third covered in peatland, but most of the once vast expanses of peatlands in the lowlands now lie under agricultural land and built development or have wasted away entirely through drainage and cultivation. Nonetheless, I am not aware of a single county across our islands that has no remaining peatland of some sort.

Over eighty-five per cent of the world's peatlands are in a near-natural state. This somewhat surprising fact is due to intact peatlands occupying vast remote areas of Canada, Alaska and Siberia, where pressure from humans has been limited. In more populated areas, such as central Europe and South East Asia, much of the peatland has been drained and cultivated.

In Britain, over eighty per cent of the remaining peatlands have been damaged by drainage for agriculture and forestry as well as burning, grazing by livestock, peat cutting and pollution. In Ireland, most of the country's extensive lowland peatlands have been stripped away by commercial peat extraction to supply peat as fuel for electricity power stations. Once a peatland is drained and the vegetation is changed or lost, the peat that has built up and been stored for millennia wastes away as it dries out, decomposes and is eroded by wind and rain.

Bogs

Bogs are peatlands where the vegetation obtains all its water from direct precipitation, in areas of high rainfall and humidity. They tend to be nutrient poor and acidic as they have no access to the minerals and salts in water from the surrounding soils because of the accumulated thickness of peat. The vegetation is composed mostly of sphagnum mosses and sedges, which are among the few plants to thrive in these acidic conditions.

The sphagnum mosses are a fascinating and important part of the peatlands' functioning. Sphagnum plants grow continuously upwards while the lower parts die but are resistant to decay. Large empty cells allow the plant to store up to twenty times its own weight in water, helping keep the bog surface saturated. The plant also creates acidic conditions, making the environment ideal for itself and less so for competing plants.

There are around a dozen different species of bog sphagnum in various tones of colour, from vibrant green and gold to deep bright red. Some species form hummocks and others grow as low-lying carpets or occupy wet hollows. These high and low features create patterns of concentric rings spreading out from the centre of the bog that become more distinctive as the peatland gets older. The thin living layer of moss on the surface of the bog, where the water table fluctuates and allows oxygen to penetrate, forms the 'acrotelm'. The underlying dead material

Blanket bog, Tynehead Fell, North Pennines

Peat erosion gully with 3 metres deep peat layer, North Pennines

that constitutes the bulk of the peat soil is called the 'catotelm'.

Blanket bogs are typically found in upland regions where precipitation – in the form of rain, sleet, snow, mist or hill fog – is highest, and where conditions are ideal they can form peat on steep ground, even on slopes with a gradient of more than 30°. As suggested by their name, blanket bogs spread out over the hilltops in an unbroken mantle of peat that can extend many miles. The largest blanket bog landscape in the UK is the Flows, sometimes referred to as the Flow Country, covering over 400,000 hectares. Blanket bogs are a distinctive and striking feature of the mountainous regions of Britain and Ireland, from Shetland in the north of Scotland to Dartmoor in South-West England, and from the west coast of Ireland to the Pennines in the north of England.

Raised bogs occupy low ground as an isolated dome like a bubble of water on a flat surface in a landscape that is now generally agricultural. Raised bogs have usually formed on top of ancient, naturally filled-in lakes that have been overgrown by fen vegetation followed by bog plant species. Through the combination of lower nutrient levels in the centre of the former lake, favouring dominance by bog-mosses, and their remarkable water-holding capacity, decomposition of the accumulating plant material is slower in the centre of the developing bog. A dome of wet peat is therefore formed that may rise several metres above the surrounding mineral ground. At the edges of the bog the gentle slope of the dome leads to the more nutrient-rich ground water. In this 'lagg' zone, forming a halo around the dome, the bog gives way to fen habitat with sedges and rushes as well as scrub woodland, known as 'carr' and containing water-tolerant tree species, such as willow and alder.

Unfortunately, few raised bog sites in Britain and Ireland still retain this lagg fen-carr feature, largely because of peat digging and agricultural activities. Today, raised bogs occur mainly in central lowland parts of Scotland and Ireland, in the Solway region and North-West England, with localised examples in the eastern regions of Scotland and northern England, though there are fragmentary remains as far south as Romney Marsh in Kent.

Fens

Fens form in areas of water collection. Consequently, they may occur in a variety of settings from relatively steep seepage zones to broad river floodplains. Most lowland examples occupy flat and gently sloping ground or low-lying depressions and receive water that has been in contact with the surrounding mineral-soil landscape. They vary in vegetation type depending on whether they are 'rich' fens, usually on more alkali soil where they overlie calcareous rocks, such as limestone, or 'poor' fens that are more acidic, but not as acidic as bogs.

Fens support a wide range of plants depending on the soil type, with the more alkali sites generally dominated by tall sedges, rushes and reeds and the acid sites having similarities to some bogs in supporting sphagnum mosses. Fens in confined shallow basins and narrow valleys are scattered throughout Britain and Ireland. Numerous small fens also occur within blanket bog landscapes wherever waters from the bogs congregate. The term 'blanket mire' incorporates this combination of bog and fen with its rich biodiversity. Extensive fens on large flat floodplains are more restricted, with the biggest examples in the East Anglia Fens, Insh Marshes in the Highlands of Scotland, Anglesey in Wales and the River Shannon in Ireland.

Wildlife

Bog wildlife

One of the main reasons that bogs are so important for nature conservation is that the extreme, wet, nutrient-poor conditions support many specialist plant and animals found nowhere else. Unfortunately, these are now some of our rarest and most rapidly declining wildlife populations. Despite bogs being the largest semi-natural habitat in Britain and Ireland, the majority have been damaged and their biodiversity is being lost at an alarming rate.

The sphagnum mosses are the keystone species in bogs, able both to thrive in and to help maintain the peat-forming system. These mosses were extensively gathered for use as wound dressings in the First World War because of their sterile and absorptive properties. The dozen or so sphagnum species that occur in bogs vary in colour, which helps in their identification. They also differ in the form that they take, either growing as hummocks up to a half a metre tall – for example red bog-moss (*Sphagnum capillifolium*), Austin's bog-moss (*Sphagnum austinii*) and the rarer rusty bog-moss (*Sphagnum fuscum*) – or as lawns with papillose bog-moss (*Sphagnum papillosum*).

A distinctive red-coloured lawn-forming moss has recently been split into two species that occur in Britain and Ireland – Magellanic bog-moss (*Sphagnum medium*) and another rarer species (*Sphagnum divinum*), which has yet to be given a common name. In Scotland, the latter species is so far only recorded from old collection specimens. Unfortunately, the site they came from has been damaged by windfarm development and forestry and the moss is no longer found there. Other sphagnum species are found submerged in pools, such as feathery bog-moss (*Sphagnum cuspidatum*) whose identifying feature is that when lifted out of the water it looks like a handful of wet fur or, as it is sometimes described, like a drowned kitten.

The low nutrient availability in bogs has led to spectacular adaptation in the form of insectivorous plants. Several species use various ingenious methods to attract and trap insect prey, which is then digested and absorbed into the plant. Charles Darwin published a book on insectivorous plants in 1875, describing his extraordinarily detailed and thorough examination of several species. In his autobiography he wrote, 'The fact that a plant should secrete, when properly excited, a fluid containing an acid and ferment, closely analogous to the digestive fluid of an animal, was certainly a remarkable discovery.'

The most noticeable and widespread insectivorous plants on bogs are the sundews (*Droseraceae*). These derive their name from the glistening halo of sticky, dew-tipped tentacles projecting out from the plant's leaves. Small insects are attracted to the sweet

Round-leaved sundew

secretions and become entrapped, eventually dying from suffocation. Sundews display 'thigmonasty', in that they are sensitive to touch and will respond by bending their tentacles towards the leaf centre, bringing the prey close to special glands that secrete dissolving enzymes. The resulting nutrient mixture is then absorbed into the plant. There are three main species found on bogs: the round-leaved or common sundew (*Drosera rotundifolia*), which grows among sphagnum mosses and is widely found in peatlands across Britain and Ireland; the oblong-leaved sundew (*Drosera intermedia*), largely confined to western areas; and the great sundew (*Drosera anglica*), which is mainly found in Scotland and western Ireland. Contrary to the indication given by its Latin name, the *Drosera anglica* is now rather rare in England. Such a strange plant has, not surprisingly, been considered in former times to possess many powers. The 16th-century herbalist John Gerard in his book *The Herball or Generall Historie of Plantes*, referred to physicians who thought:

> *'this herbe to be a rare and singular remedie for all those that be in a consumption of the lungs, and especially the distilled water thereof: for as the herbe doth keepe and hold fast the moisture and dew, and so fast, that the extreme drying heate of the sun can not consume and waste away the same; so likewise men thought that heerwith the naturall and lively heate in mens bodies is preserved and cherished'.*

Gerard noted that contrary to these beliefs, the plant had toxic qualities making it unsafe for consumption and more likely to result in early death than provide a cure. Sundew has also been known by the name 'youthwort' as it was said to stir up lustful behaviour in cows, and when distilled with wine it provided aphrodisiac and strengthening qualities for humans.

Other insectivorous plants include the common butterwort (*Pinguicula vulgaris*) which generally occurs on the margin between the bog and the many small fens that criss-cross blanket bogs. The plant's succulent glistening leaves were once rubbed on cows' udders to protect the milk and resulting butter from evil influences. Its delicate violet flowers appearing from May to July are held high up on stalks, as with the other insectivorous plants, to prevent useful pollinators becoming prey. We have much to be thankful for with all these natural midge-traps, and if that wasn't enough there is even a species that preys on the hapless insect's underwater larval stage. The bladderwort (*Utricularia vulgaris*) grows in pools and has small bladder-like traps, a few millimetres wide, that catch insect larvae and water fleas. Tiny hairs at the entrance to the trap, when triggered, cause the bladder to open and a vacuum force sucks in the prey. Early 20th century introductions of large, North American pitcher plants (*Sarracenia*) to bogs in Ireland and later to several sites in England and Scotland have been a cause for concern. These invasive plants have hollow jug-like leaves up to thirty centimetres tall, in which insects fall because of the slippery lip to the pitcher and then become trapped in the liquid contained within.

Alongside the mosses, bogs naturally hold several species of sedge. The most abundant and obvious are the cotton-grasses, whose long, thin leaves turn from green to a distinctive reddish-brown in autumn. Common cotton-grass (*Eriophorum angustifolium*), sometimes called bog cotton, grows across the wettest parts of the bogs and thrives in freshly dug or eroded peat and so provides a good guide to walkers to avoid these more treacherous parts. Its unremarkable flowers in late spring form seed heads with long white strands to aid seed dispersal.

On common cotton-grass, several flower heads occur on a single stem and resemble straggly tufts of cotton. The name *Eriophorum* is derived from the Greek meaning 'wool bearing'. In early summer it can appear as if a giant bag of cotton wool balls had been scattered over the bog in a breathtaking sight. In some parts of Scotland, the scene is likened to an army of fairies holding aloft their white standards. Another species, hare's-tail cotton-grass (*Eriophorum vaginatum*), grows in tussocks, generally on drier ground than the common cotton-grass, and has a single dense cotton tuft on each stem. The 'cotton' was used instead of feathers to make pillows in South-East England and was widely collected to make dressings for wounds in the First World War. Cotton-grass flower heads also provide an important source of protein for grouse, a group of game birds associated with peatlands.

Heather (*Calluna vulgaris*) is perhaps one of the most commonly recognised plants on bogs, although its main habitat is drier heathland. It is often called ling, has deep lilac flowers and grows in the drier areas of the bog on hummocks or drained peat. The name *Calluna* comes from the Greek 'to sweep' as it was used to make brooms. Heather was also used to provide yellow dye for textiles, including Harris tweed. Not ideally suited to wet bogs, this plant thrives in drier habitats, with its small-rolled waxy leaves designed to reduce water loss. An extensive presence of tall leggy heather indicates a bog that has dried out due to drainage or other damage.

Cross-leaved heath (*Erica tetralix*), often confused with heather, gets its name from the distinctive whorls of four leaves that occur along the stem. Found in wet parts of the bog among the sphagnum mosses, this evergreen shrub has abundant small pink bell-shaped flowers from July to September. The flowers droop to protect the pollen from the rain and are an important food source for bees. Sticky, adhesive glands on leaves, sepals and other parts of the plant prompted Charles Darwin to suggest that this species might be a partially insectivorous plant but little, if any, research has been done on this. A rare relative of the heather bog-rosemary (*Andromeda polifolia*) also enjoys wet bogs.

Harestail cotton grass with bog asphodel (*Narthecium ossifragum*)

This is a small evergreen shrub with delicate pink flowers and leaves that resemble those of the rosemary herb. It is confined to central Britain from mid-Wales to southern Scotland and central parts of Ireland.

Bog myrtle (*Myrica gale*), also known as sweet gale, is a woody deciduous small shrub with sweetly scented leaves. To cope with the nutrient-poor conditions in bogs, its roots have nodules that contain nitrogen-fixing bacteria to provide food for the plant. The astringent oils within the plant have long been used to deter biting insects and are extracted commercially today. It has been used historically and in some modern beers for flavouring as an alternative to hops, and without their sedative effect. Vikings are said to have driven themselves crazy before battle by drinking infusions of bog myrtle.

Although trees can be associated with bogs and grow on the peat in some parts of the world, in Britain and Ireland this is not the case in natural wet bogs, as there are no native tree species that can thrive in the persistently saturated and low-nutrient conditions. Trees that are seen growing on bogs invariably arise along more nutrient-rich seepage lines, flushes and stream courses or where the bog has been damaged and the soil is drier. A rare habitat known as bog woodland does contain stunted individual or small groups of trees on the wet peat but more often occurs as a transition, where bog and woodland meet and co-exist side by side, rather than one on top of the other. This atmospheric and enchanting bog woodland complex supports many rare and threatened species that benefit from the combined system.

Among the insects on bogs the most obvious and spectacular are the dragonflies. A quarter of dragonflies found in Britain and Ireland are restricted to bogs. Names like black darter (*Sympetrum danae*), white-faced darter (*Leucorrhinia dubia*), four-spotted chaser (*Libellula quadrimaculata*) and common hawker (*Aeshna juncea*) all recognise the predatory nature and superb flying skills of the adults. The larvae live in still or slow-moving water in pools and ditches within the bog and are also voracious hunters capable of eating small fish and tadpoles – though neither live in the pools of true bogs. This larval stage lasts for up to five years, but the flying adults survive for only a few weeks, during which they mate.

Another predatory insect on bogs, particularly lowland raised bogs, is the raft spider (*Dolomedes fimbriatus*). This is one of two species of raft spider found in Britain and Ireland, the other being the fen or great raft spider (*Dolomedes plantarius*), which is found in only a few locations in southern England and Wales. What makes the raft spiders so distinctive is their huge size. The body of females can be up to two centimetres long with legs spanning seven centimetres, meaning they can cover the palm of a human hand. They are rich brown-black in colour with cream stripes along the sides of the body. Their long hairy legs allow them to float on the water and move across it with ease in the hunt for insects and even small fish. The female spider carries an egg sac containing several hundred eggs for about three weeks, periodically dipping it into the water to keep the eggs

Clockwise from top left: four-spotted chaser (*Libellula quadrimaculata*), mire pill-beetle (*Curimopsis nigrita*), large red damselfly (*Pyrrhosoma nymphula*), large heath (*Coenonympha tullia*), black darter male (*Sympetrum danae*), black darter female, common hawker (*Aeshna juncea*), bog bush-cricket (*Metrioptera brachyptera*)

Raft spider

moist. Just before the eggs hatch, she builds a nursery web about twenty-five centimetres across in vegetation close to the water. Here she guards her young spiderlings for around ten days or so until they disperse into damp vegetation nearby.

The large heath butterfly (*Coenonympha tullia*) is largely confined to wet bog habitat in northern Britain, Ireland and a few isolated sites in Wales and central England. The main food plant of the larvae is hare's-tail cotton-grass, although in Yorkshire it tends to be common cotton-grass, and the adults' main nectar source is cross-leaved heath. They are a good indicator of the condition of a bog, preferring areas with plenty of sphagnum at the base of the cotton-grass tussocks. Adults always sit with their dull orange-brown wings closed, showing distinctive dark spots.

The wide, open expanse of bogs provides safe nesting habitat for many species of wading birds that can be found there from May to late July before heading to their wintering sites in lower-lying farmland and coasts. Laying their eggs in simple shallow depressions in the mossy ground or under sedge tussocks, these birds avoid nesting near trees, which provide perches for predatory birds and built structures such as wind turbines are also avoided. The summer plumage of golden plover (*Pluvialis apricaria*) is a striking mottled gold and black, but it is their haunting melodic call that is the most notable feature. Other birds, such as red-throated divers (*Gavia stellata*), that frequent small lochs within the bogs also adopt this far-reaching form of call that echoes across the moss and cotton-grass.

Less common, but even more characteristic of healthy or recovering bog, is the dunlin (*Calidris alpina*), a small wader with a black belly and slightly down-curved bill, found on blanket bogs across Britain and Ireland. Even rarer is the greenshank (*Tringa nebularia*), which during the breeding season is to be found on blanket bogs in the north and west of Scotland. These and many other birds that utilise peatlands for feeding rely on vast numbers of craneflies (*Tipula* species and *Molophilus ater*), otherwise known as 'daddy long legs', whose larvae live among the damp vegetation. Adult craneflies emerge just at the time young birds hatch, but changes in climate and damage to peatlands have created a mismatch, with fewer craneflies available for the birds to feed on at this crucial time in their early life.

In winter, a key species of the northern and western bogs is the white-fronted goose (*Anser albifrons*) with a distinctive white patch at the front of the head and bold black bars on the belly. Their breeding grounds are on peatlands in the Arctic, around Greenland and

Siberia, and they migrate to Britain and Ireland for the winter. Greenland birds can be identified by their orange bills and the Siberian race have pink bills. The Greenland white-fronted goose is most commonly associated with blanket bogs and raised bogs in the west of Scotland and in Ireland. The birds congregate on the bogs to feed on leaves and tubers of bog grasses and sedges and will also roost on small pools scattered among remote wet bogs. Around 20,000 birds make the 5,000-kilometre round-trip from west Greenland each year. Numbers have fallen by half since the turn of the century, thought largely to be due to threats to their breeding grounds.

Golden plover

The hen harrier (*Circus cyaneus*) is a bird of prey most associated with blanket bogs and raised bogs while also using grassland for foraging and heather moorland and scrub for nesting. The striking blue-grey plumage on males is unmistakeable, with the females being mostly brown apart from a distinctive white rump and barred tail. They can be seen flying and gliding low over the bog surface as they hunt for meadow pipits and voles. In winter, several pairs of hen harriers will come together to form communal roosts, often on raised bogs or fens. The name hen harrier reveals that this bird used to include young domestic fowl among its prey, but today the conflict is mainly with grouse moor managers concerned about losses of grouse chicks. Once widespread across Britain and Ireland, the hen harrier population has been severely reduced by illegal persecution, particularly in England, where only a handful of pairs nest compared with around five hundred pairs in Scotland, a hundred pairs in Ireland and around forty pairs in Wales.

Fen wildlife

Fens display an enormously rich variety, from small seepage zones in a blanket mire landscape to the vast sweep of the Insh Marshes or the Somerset Levels. They encompass habitats of open water, tall reedbeds, short sedge carpets and scrubby, wet carr woodland. Rich fens fed by alkaline waters support a large array of insects and plants, with some sites such as Wicken Fen in Cambridgeshire boasting over 8,500 species.

The tall plant communities that form extensive reedbeds are usually dominated by common reed (*Phragmites australis*), capable of growing up to four metres. The expanse of tightly packed golden stems and feathery flower spikes, seen from late summer onwards, is a distinctive feature of these reedbeds. Common reed grows by spreading underwater tubers, called rhizomes, that can extend for several metres. A stand of reeds covering over a square kilometre is usually made up of clones of the original individual, whose lineage can stretch back over a thousand years. Great fen sedge (*Cladium mariscus*), sometimes called saw sedge because of its tough, serrated leaf edges, is another tall reed-like plant that is less abundant but does form large beds. Traditionally the hollow stems of reed and the leaves and stems of sedge are used for thatching roofs, and roots were used in basket making.

Another of the iconic fen plants is the reedmace (*Typha latifolia*), with its distinctive sausage-shaped, velvety brown seed heads. Since Victorian times, the plant has confusingly been called bulrush, a name that more correctly belongs to an entirely different species of rush (*Scirpus lacustris*), which lacks the distinctive seed head. The true bulrush of biblical fame used to make a floating basket for Moses, and among whose dense stands the baby was hidden, was probably papyrus (*Cyperus papyrus*). Many European illustrations depicted the wrong species, so initiating and popularising the error. Among the reeds grow a range of tall herbs, such as the white flowers of marsh bedstraw (*Galium palustre*), meadowsweet (*Filipendula ulmaria*) and rarities such as milk parsley (*Peucedanum palustre*), alongside the vibrant colours of the purple loosestrife (*Lythrum salicaria*) and yellow loosestrife (*Lysimachia vulgaris*).

These dense, insect-rich environments provide excellent feeding and nesting sites for birds and, since it is hard to be seen among the foliage, loud singing is an advantage. Springtime among the reeds can sound like an orchestra warming up. Sedge warblers (*Acrocephalus schoenobaenus*) and reed warblers (*Acrocephalus scirpaceus*) will frequently burst forth in chattering song, accompanied by percussion from the trill of a grasshopper warbler (*Locustella naevia*) and the 'pinging' call of a bearded tit (*Panurus biarmicus*). More correctly called bearded reedling, as it is not a member of the tit family, the bearded tit males sport a long black 'Yosemite Sam' moustache. To top it all off, if you are lucky, you may hear the booming call of the bittern (*Botaurus stellaris*).

Several birds of prey utilise the fens for hunting food but the marsh harrier (*Circus aeruginosus*), the largest of the harrier species, is most at home here and nests among the reeds. Extinct in the UK by the end of the 19th century, its numbers have gradually increased after returning to eastern England in the 1970s and now its breeding range includes northern England and eastern Scotland.

Another wetland bird species that has returned to Britain after a long absence is the Eurasian or common crane (*Grus grus*). Having not bred here since the end of Henry VIII's reign in the late 16th century, a few pairs from mainland Europe began to overwinter and eventually nest among the fens of East Anglia in the early 1980s. These magnificent birds, standing over a metre tall, have a striking loud bugling call and perform a leaping, dancing display, most obvious during the breeding season. They prefer areas of shallow water and reeds free from disturbance.

A clue to just how widespread these birds once were in Britain can be found in numerous place names containing the Old English word for 'crane', *cran*, as in 'Cranfield', or the Old Norse word *tran*, as in 'Tranmere'. The loss of lowland wetlands from these areas in the 16th century was probably the main reason for the birds' demise. Conservation efforts boosted numbers with introductions of birds from Germany into the Somerset Levels and now the population is thriving with around fifty breeding pairs and over two hundred individual birds spread across Somerset, South Wales, East Anglia, Yorkshire and eastern Scotland. The Fens of East Anglia provide some of the best places for viewing crane, including the RSPB's Lakenheath Fen reserve in Suffolk and the Norfolk Wildlife Trust reserve at Hickling Broad.

Short swards of fen vegetation, referred to as fen meadows, mainly consist of the *Carex* sedges growing through carpets of brown mosses in the rich fens and sphagnum mosses in the poor fens. The short swards are maintained by light grazing from livestock. If they were not grazed they would eventually become dominated through natural successional processes by shrubs, such as willow and alder. This carr woodland is a natural feature of many fens and an important habitat, not least for birds such as nightingale (*Luscinia megarhynchos*). Managing the fen to allow the different successional stages to survive side by side in a single site is a challenge for conservation and will remain so until much larger areas of land are available for new fen to form and old fen to succumb to woodland in a natural equilibrium.

Black bog rush and saw sedge at Pollardstown Fen

Hybrid of early marsh-orchid, (*Dactylorhiza incarnata*) and southern marsh-orchid (*D. praetermissa*), Hatfield Moors

Among this, short vegetation wading birds such as snipe (*Gallinago gallinago*) and lapwing (*Vanellus vanellus*) will make their nests. A great variety of flowering plants put on a wonderful display of colour between June and August, including plants whose names give away their favoured habitat, such as marsh cinquefoil (*Potentilla palustris*), marsh lousewort (*Pedicularis palustris*), marsh marigold (*Caltha palustris*), marsh thistle (*Cirsium palustre*) and several orchid species, such as marsh orchids (*Dactylorhiza*), marsh helleborine (*Epipactis palustris*) and in a few places the fly orchid (*Ophrys insectifera*), whose flower is shaped like an insect, primarily to attract digger wasps as pollinators. The availability of calcium in the alkaline fens makes these ideal places for snails, especially whorl snails (*Vertiginidae*), several of which are threatened species. These tiny snails, often only a few millimetres in diameter, are an indicator of the health of a fen.

Fens and their wide array of flowers support large numbers of flying insects. Some fen sites in England hold over half of the UK's dragonfly species. Typical fen butterflies include marsh fritillary (*Euphydryas aurinia*), whose larvae feed on devil's bit scabious (*Succisa pratensis*), a lovely deep blue flowering plant so named because its roots end suddenly as if bitten through by the devil below. It was commonly used to treat skin mite problems including scabies; sadly, the fritillary is now extinct in the eastern half of Britain. One of the most enigmatic of fen butterflies is the swallowtail (*Papilio machaon*), our largest butterfly, with black and yellow wings spanning ten centimetres and a distinctive short tail at the end of each hind wing. Now found only in the Norfolk fens, its larval food plant is milk parsley, and the adults are especially attracted to the flowers of ragged robin (*Lychnis flos-cuculi*) growing among the reeds.

In summer the reeds are festooned with shiny silver webs made by several species of money spider. These are the smallest of our spiders, less than five millimetres in length, and there are a huge number of different species. It was believed that if they got caught in your hair it would bring good luck and wealth.

The water vole (*Arvicola amphibius*), often incorrectly referred to as a 'water rat' after the character Ratty in *The Wind in the Willows*, is actually a large vole with short ears and a short furry tail. The diet of a water vole consists mainly of vegetation and, although a

4,500 year old bog oak, Great Fen, Holme Fen

burrowing animal, it often makes aerial nests the size of a small football among the reeds and away from predators, such as mink and the true rats.

Peatland archaeology

Waterlogged, acidic and anaerobic conditions in peatlands are ideal for preserving organic material such as flesh, textiles, wood and plant matter. The archaeological resource from dry lands is considerably impoverished by comparison and largely restricted to hard materials of metal, stone and pottery. Peatland archaeology therefore provides a unique insight into people's lives and culture, stretching back thousands of years.

The peatland archive operates on different levels. Underneath the peat some sites retain a preserved record of human activity from before the rapid formation of peat, resulting from a shift to a wetter and cooler climate in a period spanning around 9,000 to 4,500 years ago, depending on the location. Ancient landscapes including woodlands, field boundaries and buildings that have been lost through damage and decay elsewhere in our modern farmland are all protected under deep oxygen-free layers of peat. One of the most extensive Stone Age monuments in the world was found in a peatland at Céide Fields on the north coast of Country Mayo in Ireland. The site contains field systems and megalithic tombs dating back 5,000 years that have been excavated and displayed as part of a major visitor centre exhibition.

One of the most abundant features preserved under the peat is the remains of trees that grew in ancient woodlands covering much of Britain and Ireland after the end of the last Ice Age over 10,000 years ago. Naturalists and historians have long remarked on the presence of branches, tree stumps and whole trees under the peat. John Leland, writing in 1543 about the Isle of Axholme in Lincolnshire, describes finding Scots pine as 'firre trees overthrown and coverid with bogge and mershe'.

Some (though by no means all) ditches or exposed peat surfaces will reveal pieces of wood thousands of years old that look as if they have only recently been placed there. At the RSPB Forsinard reserve in the Flow Country, evidence of former hazel woodland was found at the base of the peat with several perfectly preserved hazelnuts each bearing the distinctive teeth marks of voles that had extracted the seeds some 4,000 years ago.

In the 19th century there was a thriving tourist industry in selling souvenirs, jewellery and ornaments made from the dark-stained, almost petrified, remains of trees known as bog wood or bog oak from northern Europe and Ireland. Even today in Ireland, tree stumps several thousand years old can be seen stacked up in farmyards ready for sale or are used to make furniture. It is disturbing to consider that these rare artefacts, as important to archaeology and our understanding of the past as any treasure from the Egyptian tombs, can be so poorly treated.

The study of ring-spacings in the trees, termed 'dendrochronology', can tell us about growth rates and climatic conditions from past millennia, vital to understanding the impact of our own climate change period. There has been considerable debate about whether climate change or tree felling by Neolithic farmers brought about the increase in growth of peatlands to replace British and Irish forests. While humans clearly had a role, there is increasing evidence from examining peat deposits that changing climate was the most widespread cause for the woodland decline and peatland expansion.

The second level of the peatland archive is within the matrix of the peat itself. Laid down at the rate of around a millimetre per year, the peat can reach ten metres deep in some bogs, providing an amazing treasure trove of preserved information in chronological sequence. The remains of plants and animals retaining microscopic detail in the palaeoecological record provide a detailed account of the populations of individual species over thousands of years. We can see how life changed within the peatland as species responded to different conditions of wetness and dryness caused by human impact and climate. We can also see which plants grew in the vicinity of the peatland, or even further afield, through the preserved 'pollen rain' blowing in from the surrounding landscape and trapped in the peat. The peat layers can be precisely dated thanks to the presence of minute shards of glass and lava from known volcanic events as well as radio-carbon dating of preserved plant material.

Bodies in the bog

In 1983, workers mining peat at Lindow Moss in Cheshire came across part of a skull with skin tissue and hair attached. Police investigations resulted in a local man confessing to murdering his wife and disposing of her body in the bog some twenty years earlier. Analysis of the skull aged it to just under 2,000 years old, from a male living in Roman-occupied Britain. The accused man tried to revoke his confession but ended up being convicted anyway.

A year later at the same bog, another more complete 2,000-year-old body of a man was discovered by peat-mining workers. Lindow Man, or 'Pete Marsh' as he was dubbed by journalists, now resides at the British Museum in London. Although only the upper torso, head and arms were found, it is one of the best-preserved examples of a bog body discovered in Britain. The remarkable preservation properties of peatlands retained the body's external features, including hair, trimmed beard and manicured nails along with internal organs and even the stomach contents. This wealth of ancient material reveals that the man died in his mid-twenties and his last meal was probably an unleavened bread made from wheat and barley. He had consumed a drink containing mistletoe, said to be reserved for druid priests. Death was by multiple causes, suggesting a punishment killing or ritual sacrifice.

Hundreds of such bog bodies of men, women and children have been found in North-West Europe, mainly by peat cutters from the 18th century. The oldest bodies date back 8,000 years, but most are from the Roman Iron Age, around 2,000 years ago. Ireland is one of the most abundant sources of bog bodies and, as elsewhere, many seem to be accidental burials, perhaps from unwary travellers becoming trapped whereas in the later prehistoric period, more violent death with multiple injuries is a common feature. Recent research also shows that many of the Iron Age bodies, as well as body parts and other artefacts,

Cloncavan man, preserved Iron Age body with distinctive hairstyle, unearthed by peat cutting at Clonycavan, Ballivor, County Meath

have been found along modern parish boundaries with the suggestion that these follow ancient tribal boundaries. These were regarded as important places for human sacrifice and votive offerings of precious metal items and food, such as bog butter.

Studies of the bodies indicate ritual killings of high-status individuals. Among the best preserved of the Irish bog bodies is Oldcroghan Man from Clonearl Bog, County Offaly, in the Irish midlands. Dated to the early Iron Age, around 2,200 years ago, the man was exceptionally tall at six foot three inches (1.91 metres). He was in his mid-twenties, with well-manicured nails and a prestigious leather armband, indicating that he was a leader or perhaps a king. He was believed to have died from a stab wound to the chest, then to have been decapitated, dismembered and disposed of into a bog pool. There are deep cuts under the nipples – a feature shared with other bodies. Among Celtic tribes, suckling the king's nipple was a mark of respect. There is speculation that a king blamed by the community for a bad harvest or other misfortune was sacrificed to ensure better times in future. Cuts to nipples are thought to be a symbolic gesture meaning that the individual no longer could hold a high rank and were a mark of disgrace.

A similar example of ritual killing is seen with a bog body known as Clonycavan Man, found at Ballivor Bog in County Meath. Also belonging to the early Iron Age, this individual was of a slight build about five foot nine inches tall (1.76 metres) and between twenty-five and thirty-five years old. Clonycavan Man died as a result of several blows to the head, presumably from an axe. At the time of his discovery in 2003 he hit the headlines because of his impressive mane of hair, swept upwards in a mound on top of his head. Both Clonycavan Man and Oldcroghan Man are on display at the National Museum of Ireland in Dublin as part of a series of exhibits about Iron Age Ireland and its people.

Some bog bodies may just be unlucky, innocent travellers. At Gunnister in Shetland, two men in the 1950s were digging for peat and came across a body dressed in woollen clothing. Little of the body parts remained but the clothes, including a woollen shirt, breeches and long coat, were all well preserved. Among the remains was a knitted wool purse containing 17th century Scandinavian silver coins as well as personal items, including a birch stick, a horn spoon and a quill. The body appears to have been deliberately buried with the possessions carefully laid beside it. The finds are now at the National Museum of Scotland in Edinburgh.

The famous 2,000-year-old Tollund Man from Denmark had beside him a range of peat-cutting tools, demonstrating that the peatlands store information not just about the people themselves, but about their way of life and the activities they undertook. The widespread use of peat as a fuel as far back as the Bronze Age is increasingly becoming evident. On the Isle of Barra in the Outer Hebrides of Scotland, modern peat workers found a pyramidal pile of stacked peat turves a metre below the current peat surface, which was dated back over 3,000 years. Some of the turves even contained finger and thumb impressions of the ancient peat cutters.

Trackways

The most obvious signs of human activity among peatlands are the numerous wooden trackways, many dating back to the Bronze Age, with some even earlier from the Neolithic, and widely used until medieval times, before the major period of peatland drainage.

The Sweet Track in the Avalon Marshes in Somerset is one of the most famous trackways in Britain and was named after Ray Sweet who discovered it while ditch cleaning. Archaeological excavations in the area during the 1970s revealed several tracks across former bog and fen linking areas of high ground, probably containing human settlements. Some of the trackways were constructed from brushwood laid on the peat and overlain by larger branches and others were more engineered, with cross stakes driven into the ground supporting a raised platform of oak planks. Examination of the tree rings in the timbers of the two-kilometres-long Sweet Track dated it to over 5,800 years old, from the Neolithic period.

While trackways would allow ease of travel across the wet ground, they are also considered to have been widely used to allow access into the peatland for hunting game and for placing ritual offerings into the water, including human sacrifices. Finds of objects include unused and, at the time, valuable polished stone and bronze axes that had presumably been offered to the water gods. A near life-sized figure carved out of alder wood, known

Iron Age trackway dated to 2,170 years ago at Corlea, County Longford

as the Ballachulish Goddess and possibly dating to the Iron Age, was found in a bog at Ballachulish near Fort William in Scotland.

The National Museum of Scotland houses a 2,000-year-old bronze cauldron measuring forty-six centimetres in diameter from a bog at Kyleakin on Skye. Also on display is a similar aged alder wood barrel or keg, containing preserved butter. Several of these 'bog butter' finds have been made in Scotland and Ireland, and they are thought to have been ritual offerings rather than simply misplaced storage vessels.

Wooden trackways may also have performed a function as a status symbol marking tribal boundaries. The construction of a heavy timber track standing out clearly in the peatland expanse would have been an obvious show of a tribe's strength and skill. One of the largest examples in Europe of a major Iron Age trackway is the one uncovered from a bog at Corlea in County Longford, Ireland, during peat-mining works in 1984. The 2,170-year-old structure is an example of a 'corduroy road' built from split planks on top of raised rails suitable for wheeled traffic. Precise dating to 148 BCE was achieved through dendrochronological study of the timbers. The track is one kilometre long and laid with three-metre-wide oak planks. Estimates suggest it was built in a single year and would have required the felling of three hundred large oak trees. This substantial structure survived less than ten years before sinking into the peat and remaining unseen for 2,000 years. An eighteen-metre section of the track is on display at the visitor centre in Corlea with the rest preserved under what remains of the bog after peat mining ceased.

Modern wooden trackways or 'boardwalks' found on many of the peatlands now managed as nature reserves serve a dual purpose. They allow people access onto the wet ground to experience the peatland close at hand but also protect against the damaging impact of many visitors. Just as wet sand can support a person's weight unless they wriggle and sink, so a natural peatland with its layer of mosses can support a person before frequent passage converts it into a mushy soup that can take years to recover.

Preserved landscapes

The third dimension to the peatland archive is its surface. Whereas much of the British and Irish countryside has been turned to modern farmland and plantation forest, our most natural peatlands present a scene that has remained relatively unchanged for thousands of years. Looking out across the huge expanse of the Flow Country, we see a landscape rich in wildlife that would be familiar to a Neolithic farmer. With Europe having lost so much of its ancient peatland, and indeed its natural open space, protecting what little remains of these landscapes from visual intrusion must be considered alongside preserving the archaeological artefacts within and below the peat.

Lost archives

Unfortunately, so much of our peatlands have been damaged and destroyed that only a tiny proportion of their archive remains. Centuries of peat cutting across most of our peatlands has removed the most recent peat deposits, leaving only those laid down in the Bronze Age, or even earlier in some places. This is the equivalent of having an encyclopaedia with half the pages removed.

Peat cutting for fuel in the 18th and 19th centuries did reveal important archaeological finds, many now sadly lost, but it left the surrounding peatland to dry out, causing any undiscovered organic material to rot away. Commercial peat mining, with its large machinery for extracting the peat, has been an accidental mechanism of discovery by revealing several bog bodies, but in most cases such a crude approach often provides only damaged parts of the body.

Such a haphazard and destructive approach to peatland archaeology is like tearing random pages out of our ancient encyclopaedia and throwing the rest on the fire. Archaeologists would far rather the finds were left in an undisturbed bog to be preserved until less invasive technology of the future allows them to be studied without damage.

Peatlands that remain wet continue to protect the peat, but conversely, where the deposits are under modern agriculture the associated drainage removes the preserving conditions. In an average-looking arable field at Star Carr in North Yorkshire, archaeologists in the 1940s and 1950s discovered in the underlying peat evidence of a Mesolithic settlement occupied over 10,000 years ago among the fen covered shores of an ancient lake. The name *Star Carr* is from the Norse meaning 'sedge fen'. More recent archaeological work at the site has revealed some of the most exciting series of Mesolithic finds in Europe. Among the wooden houses and trackways were huge numbers of flint tools and animal bone artefacts, including twenty-one red deer antlers that had been worked to form masks or headdresses. Sadly, the agricultural drainage at the site threatens the survival of these remains and work is underway to remove them for artificial preservation.

From lost peat to lost words

There are more Scots and Irish Gaelic words connected with peat than the Inuit people use to describe snow. This reflects the long history of activities associated with peatlands, and the use of the peatlands particularly as a fuel, stretching back thousands of years. The origin of the word 'peat' itself is unclear but may derive from the Celtic *pett*, widely used to describe a piece or block of vegetable matter burned for fuel.

In Scotland 'peats' are the cut bricks, and the 'peat moss' is the place they are cut from. The Irish word 'turf' applies to the cut or about-to-be-cut material and 'peat' is the term for the unexploited areas. The Gaelic word for 'peatland' is *móin* in Irish and *mòine* in Scottish Gaelic. These words have parallels to the Welsh *mynydd* ('mountain') and come from the same Latin root as the English 'mountain', with this linkage probably the result of so much of our uplands being dominated by peatland.

The term 'moss' is derived from Germanic *mus*, meaning 'damp' or 'wet', and was originally applied to the plants, especially the sphagnum mosses, but then came to mean the places where the plants grew and were collected for use – for example, as insulation in wooden houses.

Across much of Europe all forms of peatland are called 'moors', but the meaning has shifted in northern parts of Britain influenced by the Vikings (*myrr* in Old Norse) applying only to uncultivated upland areas dominated by peatland. For lowland peatlands, the Old Norse word *mosi* meaning 'moss' was adopted in English, for example Flanders Moss in Stirlingshire. In southern parts of Britain away from Viking influence, the original meaning of 'moor' is generally retained in lowland peatland place names, such as Exmoor and Dartmoor.

Two distinct peatland types were recognised in the 16th century by John Leland, who describes *Fennes* ('fens') and *Mores* ('bogs'). *Fen* is an Old English word with Germanic origins meaning 'low-lying swampy and marshy ground'. The English term 'bog' has been taken from the Gaelic *bogach*, or *boglach* meaning 'soft, or moist, ground'. This probably relates more to the pliant saturated condition of the peat soil than the tactile delicate nature of the mosses but serves well for both aspects.

In Scotland, the Old Scots word *flowes* is used for large areas of level peatland such as Fala Flow in the Scottish Borders, the Silver Flowe in Galloway and the Flows in Caithness and Sutherland.

Peatland names are often associated with the colour red, which may come from the dramatic autumn colours of cotton-grass leaves, the wine red of the Magellanic bog-moss, or the humus-rich colour of the peat and water flowing from the bog. The red association may also be from iron ore deposits below the peat, which provided Iron Age metal workers with raw materials and red ochre dyes used as body paint by ancient Picts and other Celtic tribes, possibly providing the origin for the mischievous faerie of Irish mythology, the *fear dearg* or 'red man'.

There is a wide range of descriptive words used to name peatlands and peatland features, demonstrating an intimate understanding of the

Sphagnum mosses and bog bean (*Menyanthes trifoliata*)

natural variety of the peatland environment. Many additional words relate to traditional activities, such as peat cutting. Local historians and artists on Lewis have produced a peat glossary containing almost two hundred Gaelic words. At a conference in Ireland, I was fortunate to meet the Irish writer and documentary maker Manchán Magan, who gave a presentation in which he sought to revive Gaelic words that were falling out of use with the demise of turf cutting. Everyone in the audience was given a word and encouraged to use it and keep it alive. My words are *púcóg*, meaning 'a stock of turf set to dry', and *cróguighim*, meaning 'to put the turf on end as a footing'.

The wealth of language connected with peatlands demonstrates that for thousands of years these were a rich and important component of people's lives. Even if we are unlikely to stem the decline in ancient practices such as turf cutting, we can seek to retain the essence of peatlands as reflected in place names that often define their character as wetlands rich in wildlife. Drained, drying out and eroding peatlands do no honour to their wonderful etymology. We should also ensure that, while science and nature conservation policy rightly seek to distil down a common technical language for peatlands, we also embrace the varied language of the people whose support is essential to maintaining our peatland future. Words such as bogs, mosses, flowes and moors communicate far more than simply relying on the prosaic term 'peatlands'.

Alongside peatland language is a long history of art, music, poetry, mythology and literature. Early shamanic priests had a religious awe of peatlands as neither liquid nor solid. In Ireland, the early saints travelled to peatlands to become rejuvenated and replenished. Since then the peatlands have inspired generations of artists of all kinds. Seamus Heaney, one of the most influential poets of our times, used the imagery and horror associated with bog bodies in his collection *North* to symbolise the troubles in Northern Ireland. In one of his most widely read works, 'Digging', he also drew on his experiences growing up on a farm near Mossbawn, a bog in County Derry, where his father cut turf.

Different aspects of peatlands and peatland life have been captured on canvas in well-known works such as Thomas Wade's painting *Turf cutters*, J. M. W. Turner's etching *Peat Bog, Scotland* and in the dramatic peatland scenes of early 20th century Irish artists Paul Henry and James Humbert Craig. The photographer Fay Godwin, whose works deal with natural landscapes and environmental issues, collaborated with writer Catherine Caufield in 1991 to produce a wonderful book about the peatland at Thorne Moor in North Yorkshire. Contemporary Swansea artist Ann Jordan combined conservation and art by working with the local community to knit twelve miles of wool yarn into a blanket impregnated with heather seeds to symbolically repair an area of eroded peatland in the Brecon Beacons.

The disdain, indifference and distrust shown for peatlands is a relatively recent phenomenon of the last few centuries, stemming largely from people disconnected by time or space from those who live and work there. Through the arts there is a way to reach out to a broader audience and help more people make that emotional connection with peatlands that stretches back thousands of years.

Peatland Use

Considering that almost a third of Britain and Ireland once held peatlands it is no wonder that people have long sought to utilise the abundant, deep, organic-rich soils. The peat itself contains compressed carbon-rich plant material, an early stage in the conversion to coal, that becomes a fuel once the water is removed. Peat, as a relatively sterile light material, has been used as a soil enhancer and a material for animal bedding and packing for delicate vegetables. For thousands of years peatlands and their wildlife were a resource for hunting food and providing materials such as moss and reeds. In more recent centuries the peatlands were considered wasteland, something to be 'improved' usually by drainage, to meet modern agricultural requirements and the needs of land developers for waste dumping and built development.

Bog Iron

Bog iron was one of the earliest materials used in iron smelting. The Vikings relied on this form of iron ore and it was the predominant source of iron in the Scottish Highlands up to the end of the medieval period. Dissolved iron originating from underlying rocks or entering the bog through streams reacts in the acidic, low-oxygen conditions to form iron compounds occurring as a pan at the base of the peat. Bacterial action can also convert the compounds into nodules of iron within the peat, which were removed as a by-product of peat cutting. An iridescent oily film on the wet surface of a bog is a tell-tale sign of the presence of the bacteria and bog iron. The iron was renowned for being rust resistant, but as it only occurred in relatively small amounts its use was eventually replaced by other forms of iron ore made accessible by improved transport networks.

Peat as fuel

One of the most common uses of peat has been as a fuel for burning. Tree cover dwindled in much of the uplands during the Bronze Age through climatic change and clearance for agricultural land, denying many communities a source of fuel. The problem of keeping warm in the cold and wet climate was solved by cutting peat for burning. The calorific value is low compared with wood and it takes a lot of effort to dig and carry the heavy material, but it was better than succumbing to the harsh climate. The practice occurred in much of north and west Europe, as is evident from tools and preserved cut turves dating back two thousand years. There are also 1st-century historical references to peat cutting and to the Celtic tribes 'who burn earth'.

Peat cutting for fuel is now largely confined to remote rural parts of Ireland and the Highlands and Islands of Scotland, but for several centuries it was a common practice across much of Britain and Ireland. Major cities, such as York, Carlisle and Liverpool, were supplied with vast quantities of peat from surrounding bogs. In the 14th and 15th centuries, the fens of East Anglia supplied the city of Norwich and the cathedral priory with over 400,000 turf blocks a year. It is now known that the Norfolk Broads, once thought to be natural features, are abandoned medieval peat workings. One of the main uses of the peat from the Broads was as fuel to heat and evaporate saline water in the large-scale production of salt.

In the uplands, people had long relied on peat for fuel. Pressure on the peatlands became even more intense in the 18th and 19th centuries, as agricultural 'improvement' resulted in impoverished rural communities whose only affordable option was to cut peat to keep warm. The impact on the peatlands was severe. Extensive areas of bog on the hilltops and valley sides were drained and cut, leaving unstable saturated peat to collapse in dramatic 'bog bursts'. A whole hillside of peat could slip down the steep slopes with devastating effects on homes and farmland caught in its path.

In 1824, Emily Brontë and her sisters were caught in a heavy thunderstorm while walking near Crow Hill beside Haworth, part of the South Pennine Moors to the south-west of Keighley. Their father, the Reverend Patrick Brontë, described hearing from his house what sounded like an earthquake and headed out to the moors to look for his children. The noise was caused by a bog burst in which hundreds of tonnes of peat and water had suddenly discharged down the hill, covering fields and leaving behind a peaty crater a kilometre in diameter and four metres deep. The children were found safe and well having sheltered at nearby Ponden Hall. Emily went on to incorporate the wildness of the moors in her novel *Wuthering Heights*.

While some people speculate that bog bursts are a natural feature caused by too much water for the peat to support, there is overwhelming evidence that they are mainly associated with human disturbance, especially the digging of drains and cutting of peat blocks. Historical records show large numbers of bog bursts reported in the 18th and 19th centuries, during the period of intensive peat working and often in situations where peat had been cut on nearby steep slopes down to the underlying mineral layer.

Stacked peat turves for drying, County Mayo

The devastating power of bog bursts was witnessed more recently in 2003 at Derrybrien bog in Ireland when some 400,000 cubic metres of peat began sliding down a hillside. Over the course of several days it flowed for twenty miles along the course of the local river, blocked the main road into a village and severely polluted a water body intended as the water supply for a local town. The peatland, already damaged by conifer plantations, had been disturbed by construction work for a large windfarm prior to the burst.

By the 20th century, most of the valley-bottom peatlands and lower hill slopes in England and Wales had been fully exploited through peat cutting, leaving little if any peat soil. The less accessible peat high on the hill was uneconomic to transport and, as coal became the cheaper and more efficient fuel for rural households, the tradition of peat cutting was lost. The memory of those days remains in often found place names such as Moss Side, Moss Lane, Peat Road and Bog Road.

Peat cutting for domestic fuel using the ancient rights of 'turbary' has continued across many of Ireland's peatlands and particularly around the edges of lowland raised bogs. Cutting was carried out by hand using a traditional turf spade up until the 1980s, when the Irish government introduced grants that supported mechanised cutting using equipment easily attached to tractors.

Unfortunately, conflict between turf cutters and attempts to conserve peatlands came to a head around 2012, as the Irish government sought to implement EU rules requiring the protection of important peatlands. In the absence of a coherent peatland strategy, many farmers felt angry at the threat to their traditional rights and the story hit the national headlines. Tensions are beginning to ease as the government seeks to provide alternative peat-cutting areas away from sensitive bogs and to compensate for any losses as part of a more strategic approach to peatland conservation that engages with local communities.

The production of whisky in Scotland and poteen in Ireland has long been associated with peat. Initially used as a fuel to fire kilns during distillation, relatively small amounts are now used by some commercial distilleries where peat is burned to dry the barley in the malting process. The peat smoke releases phenols that are picked up by the grain and give a distinctive flavour to the whisky, although many distilleries now provide smokiness without using peat.

Commercial peat extraction

Until recently, Ireland was the only country apart from Russia to be employing industrial-scale removal and burning of peat to generate electricity. The Irish peat company Bord na Móna was established in 1946 with government support and with the aim of developing

Commercial peat mining on lowland raised bog, South Lanarkshire

Ireland's peatlands to provide electricity, employment and economic benefits. Over 50,000 hectares of peatland, mainly in the lowland raised bogs of the Irish midlands, have been stripped bare to supply three major electricity power stations. The company has now announced that by 2030 they will no longer use peat for electricity generation and that large areas of bog will be rehabilitated back towards peatland habitat. In less than a lifetime, Ireland's energy policy had resulted in almost a quarter of the country's raised bogs being largely depleted of peat.

The other major commercial use of peat is in providing soil conditioners and growing media for gardeners and the professional horticulture industry. The rise in popular convenience gardening, with easy-to-use, lightweight composts based on peat, started off in the 1960s and a decade later the growbags were considered an essential item for even the smallest balcony garden.

By the 1990s, new techniques in extracting peat saw a huge expansion in the production of horticultural peat. Previously, blocks of peat were cut and then crushed and sieved, with only a proportion of the bog being worked at any one time. The more cost-effective 'milling' method involves large areas of raised bogs being stripped of vegetation, drained and then the whole surface worked over by tractors with specially designed rakes to loosen the surface peat. After the peat has dried it is collected in mounds and removed by huge vacuum machines and then the process is repeated.

Claims that the peat is simply being 'harvested', or that it is sustainable as more new peat is formed in unworked peatlands than is removed by commercial mining, are unfounded. Peat does not regrow while the bogs continue to be drained and worked, and the rate of commercial extraction is over a hundred times that at which peat forms. It only takes twenty years for a peat company to reach down to peat layers formed in Roman times.

In response to conservation concerns about the damage to rare lowland raised bogs from peat extraction and mounting pressure from the gardening public, there has been an increase in the production of peat-free growing media and soil conditioners. A whole range of materials from composted green waste, wool, bark and wood fibre is now being used to provide commercially viable and very effective products. Using these materials helps protect peatlands and, in many cases, avoids waste from being dumped in landfill. With government targets for an end to the use of peat in gardening, even the main peat-producing firms are making the switch towards alternatives, recognising that there is an opportunity to develop a long-term and environmentally friendly industry.

Cultivated peatlands

European agricultural practices originated in the dry climate of the Middle East and were introduced to Britain in Neolithic times. Our predominantly wet, peaty soils had to be altered by drainage to accommodate the domestic grain species such as wheat that demanded drier soil. The Romans extended this into large-scale engineered drainage schemes involving thousands of square kilometres of low-lying fenlands around London and East Anglia.

Perhaps the most widespread and damaging impact on British and Irish peatlands was the era of agricultural improvement from the mid-17th century onwards. Among the most famous names in peatland drainage in England was Cornelius Vermuyden, a Dutch engineer employed in 1630 by the Duke of Bedford to undertake massive drainage works in the Fenlands of South-East England, much of which remains intensively farmed land today.

The first of the 'Great Drainings' began with the Old and New Bedford Rivers and the Denver Sluice – huge drainage cuts that are some of England's largest man-made features. The changes brought local protest from the 'fennish' people who feared the loss of their fishing and hunting lifestyle. One incident in 1637 saw a great many women and men going into Holme Fen in Cambridgeshire with scythes and pitchforks 'to let out the guts of anyone that should drive their fens'. Oliver Cromwell became involved, as this was his ancestral home, and he acted as an advocate for the fens people in making representations to King Charles I. After the civil war, however, Cromwell's parliament supported the Earl of Bedford and the other land improving investors. Despite riots and acts of sabotage the drainage works persisted. In the mid-17th century parliament approved the draining of the fens by the forced labour of thousands of Scots, captured at Worcester and Dunbar by Cromwell's soldiers in the Civil War along with five hundred captured Dutchmen. Plans are underway to recognise the hardship and suffering, and to commemorate those men whose story had largely been forgotten.

Things didn't go all the way of the agricultural investors, as the drained land caused the peat to subside below the height of the surrounding rivers. Windmills originally used for grinding corn were adapted to become pumps to remove the flood waters but could not operate continuously and were inefficient. By the end of the 17th century the land was once again flooded.

The great agricultural reformers of the 18th and 19th century considered peatlands to be worthless in their natural state. James Anderson, a Scottish agricultural economist, wrote *A Practical Treatise on Peat Moss* (1794), in which he states, 'moss… is considered as incapable of being employed as soil for rearing useful productions and therefore extensive mosses have in general been abandoned as barren and unproductive wastes.' He later published *A Practical Treatise on Draining Bogs and Swampy Grounds*. These modern farming methods were considered progress and the intensive production brought profit for some but was often decried by local people who saw the loss of a far more diverse harvest of wild game and plant products.

The consideration of peatlands as horrible unpleasant places was more often the view of outsider landowners and financers. Whilst conditions for those who lived on peatlands were presumably harsh at times and included high levels of diseases such as ague (presumably malaria), local people described an enriched way of life in a place where they were content. The new agriculture methods came at a huge social cost, with communities facing poverty and upheaval, but despite riots and protest the drainage of the fens continued at an accelerated pace.

Another cost still being faced by society today is the ongoing work to prevent the sunken peatlands from flooding. One of the most striking examples of subsidence can be seen at Holme Fen, where in 1848 a wooden post (later replaced by an iron post) was pushed into the peat almost seven metres down to the mineral clay layer and a cap placed on it at ground level to monitor the subsidence of the land after drainage. By 2015, the peat had dropped so much that the top of the post was left standing four metres above the surface. Since these times, the maintenance of sizable embankments along the rivers and the introduction of steam pumps powered by coal, and later electric pumps, has enabled the transformation of the wetlands into the modern agricultural land seen today.

On cultivated peatland the dry peat surface is also disturbed by ploughing, leaving a dry, friable surface that can blow away at a rate of several centimetres a year. In parts of

Rothschild's Bungalow, Woodwalton Fen, Cambridgeshire, showing peat shrinkage below sea-level.

the Cambridgeshire Fens and East Anglia, erosion has been so severe that large areas of farmland now lack peat soil with the exposed underlying gravels and clay being of no use for agriculture.

In Ireland, the growing of potatoes on peatland soils contributed to the Great Famine of the late 1840s that claimed the lives of an estimated one million people. Small farming communities had been forced into the peaty hill land to make way for cattle on the better agricultural ground. Desperate for food, they grew potatoes in poor conditions where the weak plants and wet cold environment allowed the rapid spread of the potato blight fungus (*Phytophthora infestans*) that caused the devastating failure of Ireland's potato crop.

After the Second World War, new designs in tractors and ploughs enabled drainage in even the deepest and wettest peat in lowland raised bogs and blanket bogs in Britain and Ireland. This was largely aimed at improving grazing land for livestock. The introduction of government grants for drainage, supported by European Union subsidies based on sheep numbers, led to even more intensive attempts at draining the peatlands.

Burning of the drained peatland vegetation was also deployed as a means of encouraging young, fresh grasses to provide better food for the sheep. However, the drainage of bogs seldom proved economically viable, particularly in the uplands. When the payment of grants for drainage ended, much of the peatland was simply left in a degraded state, where it continues to lose water, leading to erosion and wastage of the peat.

The cultivation of peatlands to grow commercial plantations of conifer trees began with the expansion of forestry from the 1950s and was the largest cause of lowland peatland loss up to the 1980s. The peat archive clearly shows that for the last 4,000 years of deep peat formation in Britain and Ireland, peatlands have not naturally supported trees. Most commercial tree species do not tolerate water tables close to the bog surface; therefore, to grow trees the bog must be drained.

New post-war machinery also meant that planting of blanket peat became a more realistic proposition, and from 1980 onwards, a combination of new ploughing methods, government drainage grants and tax incentives saw much more rapid expansion of forestry into the uplands, particularly in northern Scotland. Today, over ten per cent of UK peatlands are under plantation forests, and almost twenty per cent in Ireland.

The grants and tax incentives have now been withdrawn and forestry policy does not allow new planting on peatland, but we are left with often poorly growing trees that are susceptible to disease and windthrow and hardly worth the cost of harvesting and transport. They serve as a sad testament to the folly of peatland planting, yet there is still pressure for more trees on peatlands from a forestry industry struggling to meet planting targets and competing against high agricultural land prices. Bogs are seen as the cheap option.

Another perverse irony is that the need to tackle climate change is being used to justify tree planting on peatlands. It may be true that a forest can in some situations remove more carbon from the atmosphere per year than a bog achieves through sequestration by growing bog plants in the same time period, but in the long term an afforested bog loses more stored carbon from the peat than the trees can sequester, resulting in a net loss of carbon and greater climate change impact. The physical damage from draining, planting and felling of trees further threatens an immense peatland carbon store. Serious efforts to address climate change require protection and restoration of our bogs as well as planting new forests without compromising one for the other. Forests grow better, produce more timber and provide better carbon gains when planted off the bogs.

Another peatland use causing controversy in recent times is the sport of grouse shooting. Over the last two hundred years, upland estate owners have managed large areas of heather moorland to provide high numbers of red grouse (*Lagopus lagopus*) for shooting. Heather, particularly ling, naturally thrives in drier conditions away from the deep peat bogs. Centuries of drainage and manipulation have helped heather to dominate many areas of blanket bog with the subsequent loss of the sphagnum moss species. Grouse moor management that involves repeat frequent burning of the heather has a damaging effect on the moss layer and allows the heather to grow more vigorously, causing further peatland damage. Such regular or 'rotational' burning also inhibits the re-wetting and recovery of damaged peatlands, retaining them in a degraded heather dominated state.

The combined effect of grouse moor management and sheep grazing in these areas has led to some dramatic breakdowns in the peat-forming system, resulting in large, deep erosion gullies and bare expanses of peat. Some of the gullies can be as much as two metres deep and prove fatal for young grouse and lambs that become trapped by the steep sides. Increasingly, grouse moor managers are actively encouraging blanket bog to recover, reducing burning and grazing levels and blocking drains. Many consider that having healthy blanket bog in an otherwise dry heather moor benefits grouse through providing a source of water as well as insects such as craneflies for feeding their young. The flower heads of cotton-grasses in the bog habitat also provide valuable protein for adult birds.

It must be said that people are largely ignorant of the changes that have taken place in our uplands. They are essentially 'moor blind', simply seeing an expanse of purple heather-clad hills as an aesthetically pleasing sight but unaware that this has arisen through human manipulation replacing the natural mosaic of bog and heath, with resulting serious and harmful consequences.

Built development

Peatlands, long considered as wasteland, have been the out-of-the-way places for unsightly or dangerous development, including household rubbish tips and even nuclear waste. Soldiers from the Crimean War who returned to villages and towns around Lindow Moss in Cheshire with severe cases of the 'great pox', syphilis, were treated as outcasts and expelled to live in huts on the moss.

Today, the greatest development pressure on peatlands comes from windfarms, whose promoters are attracted by low land prices and remote, exposed locations providing high wind speeds. Large concrete turbine bases and connecting access roads, if constructed on deep peat, can severely damage the peatland. In addition, the operation of the turbines poses a collision risk to peatland bird species. Breeding waders have also been shown to avoid nesting near turbines, essentially being excluded from important breeding grounds.

A principal aim of windfarms is to provide low-carbon renewable energy, but studies show that a badly designed windfarm on a deep peat site can negate most of its climate benefits, due to the loss of carbon from the damaged bog. Some companies have focused on repairing previously damaged peatlands in and around the windfarm to mitigate the impact of the new development. However, as with the recent Strathy South Windfarm in the Flows of Caithness and Sutherland, objectors have made the case that, in such sensitive sites, more effective restoration work could be achieved without the damaging risks posed by the windfarm.

Peatlands' true benefits

In our enlightened times, peatlands are seen as beneficial to society in their natural state, and we understand the costly consequences of past exploitation. Protecting peatlands from damaging development is not only preserving the past but also offering a positive future as a place to celebrate the uniqueness of rich wildlife and uninterrupted space. One of their main long-term rewards is in their natural ability to inspire, rejuvenate and re-energise people, just as they did for the ancient saints in Ireland. Across Europe, the opportunity to escape modern stresses and experience such an uplifting environment is becoming more and more challenging. With the world waking up to the importance of our peatlands, switching from exploitation to helping people enjoy the natural experience while conserving peatlands must surely be the right way to treat the goose that lays the golden eggs.

With climate change now a global priority, the role of peatlands and their behaviour as long-term carbon reservoirs has been increasingly scrutinised in recent years. There has long been a popular misconception that peatlands are damaging because they release methane, a potent greenhouse gas that contributes to climate change. However, we now understand that peatlands are a huge asset in storing vast amounts of carbon and locking it up on millennial timescales. Methane is produced deep in the peat as a by-product of decay by bacteria that live in low-oxygen, waterlogged situations. In a healthy mire this methane is broken down by other bacteria in the oxygen-rich acrotelm before it can escape into the atmosphere. In damaged peatlands methane emissions may be reduced as oxygen penetrates the drying out peat, but there will be continued release from waterlogged drain-bottoms and peat cracks.

Flammable methane is often thought to be the source of the eerie lights referred to in folklore as 'will-o'-the-wisp', 'fairy lights', or 'spunkie' in the Scottish Highlands, though it may actually be another much less common gas based on the more reactive element phosphorus that is the true source. The names for these lights nevertheless all refer to some form of evil spirit holding a flaming torch or candle that draws travellers to their doom into the dangerous bog or fen.

Since methane forms in wet conditions, the draining of peatlands was thought to be a solution to halt the greenhouse gas emission. We now know that drainage results is a far greater problem through causing the release of large quantities of carbon dioxide as oxygen enters the system and allows aerobic decay of the peat. The loss of carbon dioxide in a drained bog far outweighs any reduction in methane loss and becomes a significant climate change problem.

Across the UK it is estimated that damaged peatlands release around twenty-three million tonnes of carbon dioxide annually – equivalent to over half of all the country's annual greenhouse gas reduction achievements in recent years. In other words, it negates half of

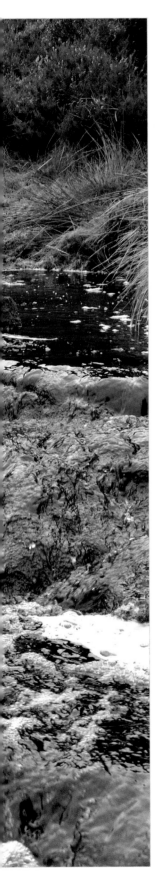

all the climate change efforts made in industry and households every year. With over three billion tonnes of carbon stored in peat deposits, we face serious consequences if peatlands are left in a deteriorating state. International climate change policy now recognises the importance of these natural carbon stores and encourages both protection and restoration to re-wet and rehabilitate the peatlands. In future, farmers could well be paid to maintain these carbon stores on behalf of the nation.

Another of the great benefits of peatlands that has been studied recently is the role they have to play in managing water. Over seventy per cent of our drinking water comes from upland peatland catchments where healthy bogs help reduce water treatment costs downstream. Conversely, damaged bogs release 'brown water' made up of particles of peat and dissolved carbon. While harmless in itself, this brown water has to be removed before the water is treated to control bacteria, as it reduces the effectiveness of filters and reacts with antibacterial treatments to form a carcinogen, which itself then has to be removed. Water companies across Britain and Ireland have now realised that it is more cost effective to treat the brown water problem at the source and prevent damage to the peatlands as well as repairing them.

The water benefits of peatlands also include flood management, through their ability to slow the release of rainwater from the hills and reduce the pressure on downstream flood defences. Drained bogs with little moss cover release storm water faster and at higher quantities with devastating and costly effect. They do so not because they absorb rainfall like a sponge, but because the moss layer adds significant 'roughness' to the bog surface and therefore slows down water movement, whereas drained or eroded bogs have multiple open channels down which flood waters can travel rapidly.

Recent major flood events at places such as the West Yorkshire market town of Hebden Bridge in the Upper Calder Valley dramatically illustrate the problems of excessive water release from the hills after storm deluges. The peatlands above the town are those frequented by the Brontë sisters, and centuries of drainage, grazing and burning have left them in a degraded state. Yorkshire Water is investing in restoring the peatlands to help retain more water in the hills for longer.

In the lowlands, a number of interesting and sustainable peatland uses are being developed. In fens, the reedbeds with their dense matrix of roots and stems provide an ideal environment for bacteria that

Damaged peatlands increase brown water in streams and rivers

Sphagnum mosses growing over felled timber plantation following peatland restoration

can help breakdown sewage wastes, releasing clean water at the other end. The reeds themselves are harvested to provide roofing thatch for houses and as a biofuel for heating and electricity generation.

With much of our lowland peatland under intensive agriculture and with evidence that drainage and cultivation are both a major source of greenhouse gases and contributors to soil loss, new approaches to wetland farming are being proposed. Having so effectively assisted the drainage of peatlands in the 17th century, we can now thank the Dutch for promoting a form of farming on peatlands where water tables remain high and peat soils are protected. Plants that thrive in these waterlogged conditions – including reeds, but also arable food crops – are grown under a system known as 'paludiculture' from the Latin *palus* meaning 'mire' or 'wetland'.

Under this system, it is also possible to harvest sphagnum mosses. The first industrial-scale trials have started in Germany, while smaller-scale trials are also underway in the UK. Sphagnum moss can be used in a variety of industrial processes as a biological filter medium. In 1915, an army surgeon, Charles Cathcart, appalled at the huge death rate from infected wounds in the early years of the First World War, turned to an ancient military practice of applying sphagnum mosses to dress wounds. During successful trials, the sphagnum was found to absorb twice as much fluid as cotton wool dressings and contained a chemical known as sphagnan that inhibited decomposing bacteria.

The use of these natural dressings became standard practice, and a major programme of sphagnum harvesting was organised across Britain and Ireland to supply the army in both World Wars. The moss is also favoured by the horticulture professionals as a better growing medium than peat. The sphagnum is essentially very young peat. Livestock production is also considered part of the paludiculture approach, where controlled grazing is used to help retain swards of short poor-fen plants of conservation importance.

Such ways of cultivating peatlands, while being sustainable in protecting the peat soils, can still be an intrusive activity affecting peatland wildlife. Paludiculture is therefore not a panacea for all peatlands but should play an important part in helping maintain farming incomes and in supporting rural communities where peatlands have been intensively damaged for centuries. It also offers the potential to provide profitable forms of land management actively requiring water regimes that are sympathetic to and supportive of the needs of adjacent wetland conservation sites. Alongside this wetland farming, conservation organisations are developing large-scale restoration schemes to return the natural diversity of peatland wildlife, such as in the Great Fen project covering almost 4,000 hectares in eastern England.

A New Era for Peatlands

In the second half of the 20th century, peatland conservation focused on halting damaging developments. Botanist and television presenter David Bellamy often hit the headlines as he campaigned against proposals to drain and cut peatlands in Ireland and Scotland. Thorne and Hatfield Moors were a focus for campaigning against the mining of peat for the horticulture industry and the Flow Country was the bitter battleground in the fight to halt conifer planting on peatlands.

National partnerships were formed to co-ordinate the efforts of conservation bodies and to work with government agencies. In Ireland during the 1970s, the Dutch – again making up for past damage to peatlands – formed the Dutch Foundation for the Conservation of Irish Bogs. Funds were raised in the Netherlands to help buy several peatlands, which were gifted to the Irish nation. A decade later the Irish Peatland Conservation Council was established to campaign for peatlands and now continues that work as well as holding important inventory details on the state of Irish peatlands.

From the 1970s, international conservation obligations, such as the Ramsar Convention on Wetlands and the EU Habitats and Species Directives, sought the protection of peatlands through designation aimed at stopping new damaging developments. National governments, however, had little funding or appetite to deal with the agricultural damage from 'perverse' subsidies and grants that were encouraging farmers and foresters to damage the peatlands.

As conservation concerns increased, various government peatland schemes were introduced in the 21st century to pay land managers to reduce burning, reduce grazing livestock and help repair the peatlands by blocking old drainage ditches. With over eighty per cent of Britain and Ireland's peatlands in a degraded state, the grants were insufficient to address the full scale of the problem. They were also of limited interest to those who perceived that more money could be made from continued agriculture. Developing a new approach to farming support is being considered as part of the future policy for the UK outside the European Union. The opportunity to align different funding streams in a way that enables peatland conservation is being promoted by politicians as a positive part of the new agriculture support system.

The IUCN UK PP partnership has sought to build consensus since 2009, the partnership has sought to build consensus on the scientific evidence around peatland benefits and the impacts of different activities. A strategic plan for UK peatlands aimed at delivering peatland restoration targets through to 2040 has been published and is supported by government-led plans at country level, all marking an important new chapter for our peatlands.

One of the cornerstones of the IUCN UK PP work was being able to build on the successes in peatland restoration that had taken place in the preceding twenty years. The early 21st century saw major advances in the methods and extent of peatland restoration. Initial projects, blocking drains on a few raised bogs, were being replicated across Britain and

Peatland restoration using a wooden peat dam to block erosion gully at Kinder Scout

Before (left) and after (right) peatland restoration on the Pennine Way at Black Hill in the Peak District.

Ireland. Some of the earliest examples were carried out by wildlife conservation charities, such as the RSPB and the various Wildlife Trusts.

In some places, local communities took the initiative and organised work parties and fundraising groups to help conserve their neighbourhood peatland. At Abbeyleix Bog in County Laois, Ireland, residents of the local town formed an action group in 2000 to protect the raised bog from commercial peat extraction by Bord na Móna. In 2010, the company agreed to lease the site to the group and is assisting in conserving and restoring the habitat.

In 2006, the local community formally constituted Friends of Langlands Moss to help conserve a local authority-owned lowland raised bog on the edge of East Kilbride, a new town in South Lanarkshire. Using government grants and money from lottery awards, the community began blocking old drainage ditches and constructing boardwalks and footpaths to allow local access. Among the benefits of this work was the sense of community with volunteers working in evenings and weekends in what many considered as a natural gymnasium, improving fitness and building social contacts. Some of the children growing up and joining in with the work gained an interest in peatlands that led to environmental careers.

As experience grew, ever more ambitious, large-scale peatland restoration projects covering hundreds of hectares began to emerge. One of the first major initiatives was in 1992 with a partnership between Fermanagh District Council and the RSPB in Northern Ireland funded by the EU to tackle the cut-over blanket bog above the Marble Arch Caves in the Cuilcagh Mountain Park. Blanket bog damaged by mechanical peat cutting and drainage was causing water to flood the important limestone caves, a major tourist attraction. Twenty-five years on, the site is now managed as a major peatland visitor attraction and demonstration site for peatland restoration.

The late 1990s saw the establishment of a peatland initiative in the south-west of England. The Exmoor Mires partnership was formed in response to concerns about the impact

that drained and cut-over peatlands were having on the river systems. In 2006, South West Water joined the initiative and injected funding to help restore blanket bog. The initiative is still running today and is monitoring the changes that are bringing improvements to water quality and to the water-holding capacity of the peatlands, as well as climate change and wildlife benefits.

Another EU-funded project, the Flow Country Partnership, was formed in 2001 following the successful campaign against tree planting, and began the task of removing trees and blocking drainage ditches over a huge area of around 15,000 hectares of blanket bog. Such ambitious achievement should be seen in the context of the overall site, which is 400,000 hectares and the largest Atlantic blanket bog in the world. In 2014, the project was extended under the banner 'Flows to the Future', with the Peatland Partnership planning to restore another 2,000 hectares of blanket bog.

In northern England, the moorlands of the Peak District and the South Pennines are renowned as important tourist areas, but in the late 1990s the peatlands were in a dire state. Years of drainage, burning and grazing combined with industrial pollution from the nearby cities had left the hilltops bare of vegetation in what can only be described as a moonscape.

In 2003, the Moors for the Future Partnership was formed to bring different interested parties together to address the problem through raising awareness, restoring the peatlands and scientific work around peatland management. The incredible achievements of the partnership are visible today to anyone walking the Pennine Way across the Bleaklow summit, where once bare peat is now covered in cotton-grass with sphagnum beginning to recover.

In 2005, the United Utilities water company partnered with the RSPB and local farmers in a project to restore and enhance the water catchments in North-West England. The work, under the umbrella name of the Sustainable Catchment Management Plan (SCaMP), included restoration of over 5,500 hectares of blanket bog.

Large-scale projects have also been established in the North Pennines Area of Outstanding Natural Beauty as well as in Yorkshire and across the Humberhead Levels covering Thorne and Hatfield Moors. Work has expanded into the lowland fens, with the Anglesey Fens project in Wales and initiatives in the Somerset Levels and East Anglia seeking

Peatland restoration using coir matting rolls and stone dams, Tynehead Fell, North Pennines.

to return rapidly depleting agricultural soils back into wet, stable peatlands, providing benefits for nature and the opportunity for new crops.

Peatland restoration is a long process. After all, it took thousands of years to form the deep peat. What is remarkable is that rapid improvements can be achieved from re-wetting a site with an almost immediate reduction in carbon loss to the atmosphere and noticeable improvements in water colour and flood water release. Peatland restoration should be seen as a long-term trajectory, starting with stabilising the damaged system, removing the adverse impacts and gradually allowing the reinstating of peatland vegetation and other wildlife.

In a world where severe climate change is inevitable, thanks to our slow response to the crisis, and where the best we can aim for is to avoid the most harmful scenarios, there have been claims that British and Irish peatlands are doomed under the warmer, drier conditions. But peatland experts have pointed to the fact that peatlands grow in almost every country in the world and that the archive in the peat store shows that during previous periods of climatic change the peatland adjusted to the new conditions and carried on supporting peat.

Of the many sphagnum varieties, some prefer wetter conditions while others thrive in drier situations. As the climate changes the abundance of different moss species adjusts to suit the new conditions. This all assumes that the bog is in a healthy state to start with and supports a good range of mosses. Damaged bogs are compromised in their ability to deal with a changing climate and the sooner we restore them the better. Even if the claims that it is too late to fully restore these damaged sites were correct, re-wetting them will at least slow the loss of carbon from the huge peat store and avoid exacerbating the climate change problem.

To ensure we no longer slip back into the past where peatlands were considered worthless it is important that we make people aware of the benefits of healthy peatlands and support those who manage them in recognition of the services peatlands provide for society. Entering the third decade of the 20th century we face some of the greatest economic, environmental and health crises in living memory, and as governments plan for the future concepts such as green recovery and nature-based solutions are being given serious consideration. It is realised that many of our problems arise from unsustainable and environmentally damaging lifestyles and that now is the time to forge a new economy that works with and benefits from a healthy and recovering environment. Peatlands are a prime example of nature providing solutions to help tackle climate change while supporting wildlife and offering health improvements, employment and cost savings for society.

Part of this new era is making sure people are properly acquainted with peatlands through enjoying a trip to the moss, the boglands or the fens. We must not let controversies over the rights of conservationists, sheep farmers, grouse moor owners, foresters and developers dominate the agenda. The more deep-rooted and established connection with our peatlands is the link between people and the rich historical, cultural and artistic appreciation of these places and their wildlife. It is within our grasp to manage peatlands in a way that retains their natural features and to adjust or reposition more invasive activity, we will be able to pass on something beneficial to the next generation, rather than the costly liability of a degraded peatland landscape.

Visiting peatlands

Britain and Ireland are fortunate in having peatlands on our doorsteps, from the built-up conurbations of Manchester, Glasgow, Cardiff, Dublin and Belfast to the most remote rural reaches of our islands.

People from all walks of life have long found solace among peatlands. Open, airy vistas provide relief from the clutter and oppression of our bustling cities and hectic work. The therapeutic benefits increase the more relaxed the visitor becomes. The initial empty

Resting place on trail at Flanders Moss

landscape gradually reveals its secrets as you become attuned to the distant calls of birds and the marvellous intricate detail of the varied plant and insect life. The best way to appreciate a peatland is to take the time to be still and let the senses gradually focus in on the natural wonders.

Visiting peatlands can be done at any time of year and every season has its own highlights. What appeals to some is the noise of the breeding season in spring and summer when the bogs and fens are alive with birdsong. For others it is the spectacle of autumn with its reds, browns and yellows, but even winter has its own dramatic rewards as the watery landscapes freeze into spectacular displays.

Peatland access does require careful management to avoid disturbing sensitive wildlife associated with the solitude and safety that these wetlands naturally provide. The plants that make up the delicate peat surface are vulnerable to trampling, especially in peatbogs where footprints and tyre tracks can remain visible for many years. They can also be dangerous places for the unwary, with deep water and saturated peat soils. The best way to be safe and avoid harming wildlife, or being harmed, is to use the paths and boardwalks provided by the visitor trails on the many peatland nature reserves and parks.

Experiencing peatlands is not just for the tourist or day-tripper but is increasingly a regular part of people's lives where communities become involved in the repair and maintenance of peatlands as a social activity. Becoming involved with a local 'Friends of the Bog' initiative or joining the local nature conservation members group provides real health and well-being benefits. Engaging in citizen science activities, helping gather information on peatland wildlife and changes in the landscape, is an activity available to all ages and is incredibly rewarding.

A visit to a peatland should involve more than just the bog or fen itself but should also take in the cultural aspects. Neighbouring monasteries, other historic buildings and archaeological features all tie into the history that has helped shape and explain the connections between people and peatlands over the millennia. Even the tracks and trails to and through the peatlands have their own story to tell. There is a real sense of wonder in following the routes of old railway lines and bog roads that once carried peat for fuel, military roads from the era of our last civil wars, and ancient

Claggan Mountain, boardwalk trail at Ballycroy, County Mayo

Bog Road trail, Connemara National Park, County Galway

routes where cattle were walked to market or where primitive religious ceremonies took place.

As with any visit to the countryside, care should be taken over access rights, and particular care should be taken for other users during sensitive times for those managing livestock, game and wildlife. The Scottish Outdoor Access Code and the Countryside Codes in England, Wales, Northern Ireland and the Republic of Ireland describe these issues in more detail.

Wetland habitats harbour biting flies, such as midges and horse flies or 'clegs', which are most prevalent in the summer months. Insect repellent and netting are the best way to avoid irritation. A more serious problem are the blood-sucking ticks, which can transmit Lyme's disease and are usually found in heather. As well as insect repellents, thick socks and gaiters help deter them. NatureScot provides further information on how to deal with tick bites and reduce the risk of infection.

Using this guide

The Peatlands given in this guide and their boundaries have been identified using the following inventories.

Scotland, England, Wales, Northern Ireland – Land Cover 2019 (UKCEH)
Republic of Ireland – Corine Land Cover 2019 (EEA)

With so many peatlands to choose from, this has inevitably become a personal selection of some of my favourite sites, while also attempting to provide a broad representation of peatland types across Britain and Ireland. The sites given are also those that are the most accessible and generally have good visitor interpretation facilities. The maps provided are intended as an aid to planning a visit and should not be used for navigation. The relevant Ordnance Survey (OS) map sheet is given at the start of each site account. For peatlands in Northern Ireland and the Republic of Ireland, the Irish grid reference system is used.

Key to Maps

- Peatland habitats (including raised bog, blanket bog and fen)
- Woodland
- – – Peatland trails given in the text
- – – Suggested connecting routes

NB these maps are provided simply as a guide to planning routes and should not be used for navigation.

SCOTLAND

Almost a fifth of the Scottish land area is peatland, making this among the highest percentage peat cover for any country in the world. Scotland holds an estimated thirteen per cent of the world's blanket bog. The Flows and the Lewis peatlands are of global importance as some of the largest and best-condition examples of this peatland type.

Blanket bog occurs widely across the hills and mountains of north, west, central and southern Scotland, with raised bogs more abundant in the low-lying Central Belt and east coast. There are relatively few large fens remaining, although Insh Marshes, within the River Spey catchment, is one of the largest floodplain fens in Europe.

Blanket bogs have been utilised for thousands of years for peat cutting, mainly for fuel, and the practice can still occasionally be seen on crofting land in remote northern and island communities. Much of the blanket bog is used for livestock grazing, mostly sheep and red deer, along with grouse shooting. Large areas have been drained in the past for forestry and agriculture. Windfarms are a notable new feature on and around both blanket bog and raised bog. Commercial peat mining for horticulture use still takes place on several lowland raised bogs, although new permissions are unlikely to be approved by the Scottish Government.

A major programme of peatland restoration has been put in place through the NatureScot Peatland Action project. Since 2012, over 15,000 hectares have been put on the road to recovery with funding provided by the Scottish Government.

With peatlands a strong part of the cultural identity of Scotland, and with a climate that allows peatlands to thrive, this is the place to experience peatlands on a grand scale with prime examples of the habitat and its full wealth of wildlife. It is therefore no surprise that much of Scotland's peatlands are protected by wildlife law and make up some of the largest nature reserves owned and managed by wildlife conservation organisations. Scotland's two national parks, the Cairngorms National Park and the Loch Lomond and the Trossachs National Park, both hold major peatland areas and are supporting local land managers in maintaining them as well as helping visitors to appreciate and enjoy them.

Flanders Moss, raised bog, Stirlingshire

NORTH-WEST HIGHLANDS AND NORTHERN ISLES

With the John o' Groats landmark, and mainland Britain's northernmost point at Dunnet Head, this region is often considered as being at the periphery. By contrast, in ancient times the counties of Caithness and Sutherland and the Orkney and Shetland islands were a Neolithic focal point, with people congregating here from across Europe. Today this region is again a centre of attention, as supporting the world's best example of blanket bog and offering a unique visitor experience among a vast peatland landscape unrivalled for its wildlife.

The North-West Highlands geological region is located in the top third of Scotland, separated from the Grampian Mountains by the Great Glen. Blanket bogs are found throughout the region along with striking examples of raised bogs at Claish Moss south of Loch Shiel and neighbouring Kentra Moss opening onto Kentra Bay in the remote Ardnamurchan peninsula. Claish Moss is largely inaccessible but there is a route around Kentra Moss utilising part of the Highland Council core path network between the villages of Acharacle and Arivegaig.

In the far north, the Flows of Caithness and Sutherland are a dominant feature displaying all the characteristics of an upland environment but with the blanket bogs largely close to sea level. This is a remote area with the human populations concentrated in the main towns of Thurso and Wick and in numerous small villages dotted along the coastline. The 16th-century cartographer Timothy Pont hailed from Dunnet in Caithness, where his father was minister – a role later taken up by Timothy himself. Between 1584 and 1596, he toured Scotland on foot, mapping the country in harsh circumstances, often being robbed and at a time when bubonic plague was rife. Sadly, he couldn't get a publisher to support him and it was only after his death that his maps were incorporated into Blaeu's *Atlas Maior* of 1665. Dunnet church displays a plaque commemorating his memory.

The influence of the Gulf Stream means relatively mild winters and often unfrozen wetlands, providing excellent conditions for the huge numbers of wading birds and wildfowl that breed and winter here. Major roads hug the coast, leaving often single-track routes to reach the interior. The Inverness–Wick railway line between Helmsdale and Thurso traverses the peatlands offering spectacular views. Ferry links connect to the islands of Orkney and Shetland, where there are several fine examples of blanket bog to explore among the dramatic coastal scenery.

Viking connections are strongly retained in this region, following the Norse arrival and settlement from the 10th century. Their legacy is found in numerous place names including

Caithness, derived from the Old Norse for 'headland of the Catt tribe', a reference to Pictish predecessors. Wonderful examples of Neolithic buildings and structures remain as testimony to the area's huge social significance and the incredible engineering skills of people who lived here between 2,000 and 5,000 years ago. After the end of the last Ice Age, Neolithic settlers would have seen a landscape across the peatlands much as it is today. There have, however, been some notable changes to the peatlands through centuries of peat cutting for fuel, livestock grazing, burning and drainage, along with the more recent incursions of conifer plantations and windfarms.

Orkney; view from summit of Trumland Hill, Rousay looking to Egilsay and Eday

SHETLAND

Map: OS 1:50,000 Sheet Nos 1, 2
Peatland Grid ref: Yell (HU 490 886)
Access point: Hermaness (HP 612 149), Mires of Funzie, Fetlar (HU 655 900)

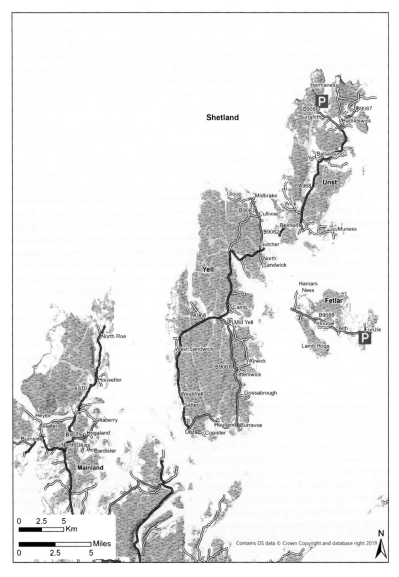

Far out from the Scottish mainland in the divide between the Atlantic Ocean and the North Sea, the Shetland archipelago shares an even stronger Norse influence than its

Hermaness National Nature Reserve boardwalk

neighbour, Orkney. It also contains a wealth of Neolithic and Iron Age structures made from stone as woodland has long been scarce. Over half the islands are covered in peat but much of this has been severely damaged by centuries of peat cutting, sheep grazing and drainage, leaving the peatland to crack and erode with huge, exposed gullies. In the less accessible parts of North Mainland at North Roe and Tingon, and on the smaller islands such as Yell, good examples of blanket bog habitat can be found along with incredible numbers of wading birds, red-throated divers and hen harriers.

The peatlands on Shetland are important breeding grounds for both the great skua (*Stercorarius skua*) known locally as 'bonxies', from an Old Norse word *bunki*, meaning 'dumpy', referring to its heavy-set build, and its smaller cousin the Arctic skua (*Stercorarius parasiticus*), known locally as 'Scootie Allan'. Another important Shetland bird species is the whimbrel (*Numenius phaeopus*). With similar features to their larger cousin the curlew, whimbrel can be distinguished by a pale stripe through the centre of the crown and by the evocative rippling whistle and trilling call from which it gets its name.

Yell is Shetland's second-largest island and is well worth exploring. Predominantly peat covered, Yell was proposed as a location for a peat-burning electricity power station in the 1970s. At the south-east of the island in Burravoe, the 17th-century merchant's house, the Old Haa of Brough, has been converted to a museum and visitor centre with exhibits describing the natural and cultural history of the island. Outside is a memorial to one of Yell's famous inhabitants, Bobby Tulloch, a naturalist, photographer, writer and great storyteller. He was the RSPB's first representative on Shetland and frequently gave captivating talks, including ones I was fortunate to attend as a student at the University of Aberdeen and again while working with the RSPB. Retiring early due to ill health in 1985, Bobby sadly died at the age of sixty-seven in 1996. One of Bobby's passionate interests was otters, which are abundant on Yell and during the long daylight hours can frequently be seen along the margins of peatland and coast.

The island of Fetlar is a major attraction for birdwatchers, offering a range of coastal and peatland species around the Lamb Hoga peninsula. Much of the eastern part of the island lacks deep peat due to the underlying base-rich serpentine rock. Perhaps the bird species most people come to see is the red-necked phalarope (*Phalaropus lobatus*). This small, delicate bird has the unusual ability for a wader of being able to swim in open water due to its lobed toes. The RSPB Mires of Funzie reserve holds over half of the UK population of fifty breeding pairs, with a hide offering great views of these charismatic birds.

Unst, the northernmost inhabited island in the UK, holds a special peatland treasure, the Hermaness National Nature Reserve, which is well worth the journey. A long boardwalk gives access to large open swathes of blanket bog bejewelled with small lochans and flanked on the west by dramatic sea cliffs. This is a birdwatching paradise with golden plover, dunlin and red-throated divers on the bogs and puffins, auks and

Bog pool in blanket bog at Hermaness

Blanket bog with cotton grass at Hermaness

gannets along the coast with even the possibility of seeing killer whales or dolphins offshore.

The long history of peatland use and exploitation on Shetland has left a challenging legacy. In extreme cases, the drying out peat becomes unstable and prone to bog bursts, which occur frequently and can be dangerous. At Channerwick in 2003, a two-hundred-metre section of the A970 road was inundated by a bog burst, with another at Uradale on Scalloway in 2012 which flooded some residences. Other events on hillsides on Mainland in 2010 at Dury Voe, and again in 2015 at Mid Kame near Voe, have caused local concerns as the areas are planned for windfarm development, which could cause further disruption to the vulnerable peatlands.

NatureScot, through its Peatland Action Scheme, has been repairing some of the eroded peatlands to halt drainage and revegetate bare peat surfaces. The remoteness of the islands means that the cost of shipping geotextiles and other materials used in peatland restoration is prohibitively expensive. Innovative solutions have been found by utilising recycled and reclaimed materials from the aquaculture industry.

At Sandy Loch Reservoir near Lerwick, Scottish Water has identified that the surrounding damaged blanket bogs are impacting on the provision of drinking water. Peatland restoration is being used to help tackle the problem.

Travel notes

Shetland can be reached either by ferry from Orkney or direct from Aberdeen to Lerwick.

Cycling is a great way to travel among the peatlands on Shetland. There is bicycle hire and repair in Lerwick. The North Sea Cycle Route, crossing both Orkney and Shetland, is part of an international 6,000-kilometre chain that is helping promote cycling in the region. National Cycle Route 1 is well marked and follows rural roads from Lerwick on mainland Shetland to the ferry crossing at Toft over to Ulsta on Yell, then continuing on the A968 to Gutcher for the ferries to either Hamars Ness on Fetlar or Belmont on Unst. There is also a public bus service from Lerwick to Gutcher.

The Tingon peninsula at Northmavine offers blanket bog expanses and views of Ronas Hill, Shetland's highest point. From the A970, take the B9078 heading west and follow signs to Eshaness. At Braewick turn north onto a small road to Hamnavoe then after 1.5 kilometres turn right onto a track with a wooden sign to Tingon (HU 243 806). Although vehicles can use this track, walking or cycling gives a better experience of the surrounding blanket bog. It is important to stay on the track during the breeding season to avoid disturbing nesting birds. The track ends at a farmhouse (HU 249 829). From here there the Atlantic coast can be visited by following the Burn of Tingon to the dramatic coastal scenery. The ruined crofts in this area are a reminder that the area held fourteen families and almost a hundred people in the late 19th century before the Clearances in 1865, when people were evicted to make way for sheep farming.

For Hermaness National Nature Reserve on Unst, leave the ferry terminal at Belmont and head sixteen kilometres north through Baltasound and turn left just before Haroldswick onto the B9086 towards the village of Burrafirth and Hermaness. At the fork in the road continue straight on to a car park and the start of the reserve on the west side of Burra Firth (HP 612 149). There is a bus service from Lerwick to Haroldswick which requires an overnight stay to make a worthwhile trip to the reserve. A gravel track and boardwalk lead onto the peatland and the spectacular west coast sea cliffs with breeding puffins and gannets.

To reach the RSPB Mires of Funzie reserve on Fetlar from the ferry terminal at Hamars Ness, take the road east for ten kilometres to a small car park at the west of the Loch of Funzie. Walk east for one hundred metres and follow signs to the hide, which is situated about three hundred metres from the road.

Shetland is well supplied by local wildlife tour operators who can provide detailed knowledge of the rich natural heritage among the Shetland archipelago of over one hundred islands.

Place Names

Shetland: the island's Old Norse name is *Hjaltland* with *hjalt*, meaning 'sword hilt' referring to the shape of the island.

Yell: Proto-Norse name *Jala* or *Jela* may have meant 'white island', referring to the beaches, or Old Norse *Gjall* meaning 'barren' but possibly of unknown Celtic or pre-Celtic origin.

Fetlar: possibly Old Norse for 'shoulder straps' relating to the Mesolithic Funzie dyke that divides the island, or derived from an older unknown Celtic name.

Funzie: from the Old Norse name *Finns* given to the legendary pre-Norse, Pictish inhabitants.

Burravoe: derived from the Old Norse *Borgavágr*, meaning 'bay of the broch' (a prehistoric circular stone tower).

Tingon: Old Norse *tunga* meaning 'tongue of land'.

Northmavine: Old Norse meaning 'north of the narrow isthmus'

ORKNEY

Map: OS 1:50,000 Sheet Nos 5, 6, 7
Peatland Grid ref: HY 358 210
Access point: Birsay Reserve (HY 340 240), Cottascarth and Rendall Moss (HY 369 195) and Hoy (HY 233 037)

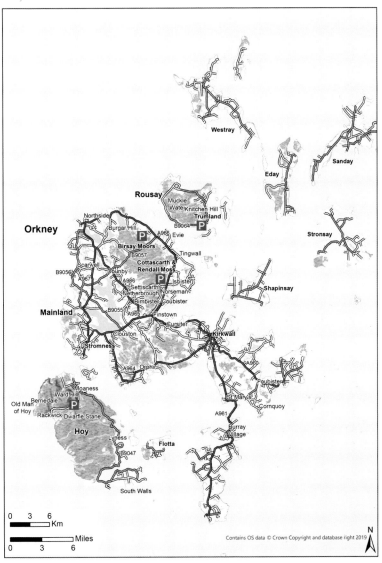

Orkney is an archipelago situated just sixteen kilometres north of the Caithness coast but separated from it by the fearsome Pentland Firth. The largest of the seventy islands

View up the Post Road glen towards the Cuilags on Hoy

that make up Orkney is known as Mainland. The name Orkney is thought to be derived from a Pictish or even earlier Neolithic tribal name, *Orc*, meaning 'young pig' and later adapted by the Norse settlers from the 9th century into *Orkneyjar*, which translates as 'island of the seals'. The islands were under Norse rule until they were given back to the Scottish Crown in 1472 and there remains a strong Scandinavian heritage, reflected in many place names. The largest areas of blanket bog are found at the Birsay Moors on the western mainland and on the islands of Hoy, Rousay and Eday.

The islands of Orkney are renowned for their wildlife and archaeological richness. The combination of stunning beaches, sea cliffs, peatland and mountains offers breathtaking scenery. Strong winds batter the islands and there is frequent cloud and rain helping make this landscape ideal for peat formation. There are still plenty of clear, sunny days and so much to see despite periods of fast-moving weather.

This is easily the best place in Britain to see hen harriers, with around eighty breeding pairs recorded in the 2016 national survey but now down to around fifty pairs. The reasons for the decline are unclear but may be linked to limited food availability associated with habitat deterioration. One of the hen harrier's favourite food species is the Orkney vole

Wilderness Track and peat cutting on Birsay Moors, Orkney

Eddie Balfour Hide, Cottascarth, Orkney

(*Microtus arvalis orcadensis*), a subspecies of the smaller European common vole (*Microtus arvalis*) and found nowhere else in Britain. DNA tests show their closest living relatives are in Belgium. Orkney vole bones have been discovered among archaeological sites dating back 5,000 years. The animals are thought to have been brought by immigrating Neolithic farmers from Europe, possibly as stowaways in animal fodder or bedding.

One of the largest of the peatlands is on the western Mainland with the RSPB Birsay Moors Nature Reserve at its western edge. The reserve covers 2,350 hectares, with heathland and blanket bog fringed by grassland and wetland. Hen harriers breed here and in winter form one of the largest communal roosts of this species in Britain, although numbers have halved in the last fifteen years. Bog pools and lochans scattered throughout the peatlands provide nesting sites for red-throated divers with their distinctive red throat patches and slightly upturned bills. Their long, plaintive cries are an intimate part of the bog experience and are believed to herald wet weather in common lore.

The dramatic landscape in this part of Orkney holds the Heart of Neolithic Orkney UNESCO World Heritage Site with standing stones, burial chambers and the 4,500-year-old ceremonial centre at the Ness of Brodgar. There is also a windfarm adjacent to the moorland on Burgar Hill, which was located and designed away from the deep peat and positioned to avoid bird collisions with the turbine blades. The peatlands have extensive areas of peat cutting dug out by hand for centuries and still used as fuel by people on the islands.

At the eastern side of the moorland the RSPB also have a small reserve with pockets of blanket bog at Cottascarth and Rendall Moss, an excellent place to see and hear curlews during the breeding season. The hide here is named after Eddie Balfour, known for his work studying hen harriers and helping in their conservation, and it remains an excellent viewpoint for their 'sky-dancing' displays and food passes between males and females in the breeding season.

Hoy, the 'high island', is the second largest of the islands and supports the largest RSPB nature reserve in Orkney with over 3,000 hectares of heathland, blanket bog and Britain's most northerly ancient woodland, Berriedale, composed of downy birch, hazel, aspen, rowan and willow. The reserve also contains Ward Hill, which at 479 metres is the highest point on Orkney.

Along with the usual peatland birds, Hoy has Britain's largest population of great skuas. Up to 1,340 pairs of great skuas nest on the island, comprising almost ten per cent of the world population. Nesting among the peatlands of Hoy, the great skua is a sea-going species that migrates to wintering grounds off the coasts of Spain and West Africa. Described as a 'pirate' or klepto-parasite, it attacks and harasses other seabirds in flight to make them disgorge their food, which it steals. Bonxies are very active in defending their nests and young, and will dive-bomb if you get too close. Holding your hand or a stick above your head while you move away will avoid any risk of a slap on the head with their feet.

Since 2015, a pair of white-tailed eagles (*Haliaeetus albicilla*), also known as sea eagles, have bred above the Dwarfie Stane, a unique Neolithic cut-rock stone on the Dwarfie Hammars. In doing so, they became Scotland's one hundredth breeding pair.

Travel Notes

The Orkney bus connects Inverness with the John o' Groats ferry and another bus on the Orkney Mainland. Trains to Thurso are connected by short bus routes to Scrabster and Gill's Bay for ferries to the island. There is also a major ferry crossing from Aberdeen.

The RSPB Birsay reserve on the mainland (HY 340 240) is best viewed from the Hillside Road (B9057) between Evie and Dounby, and from the Birsay Moors hide on Burgar Hill. To get there, follow the brown tourist signs from the A966 north of Evie at HY 357 266. Cycle Route 1 of the National Cycle Network runs through Evie, three kilometres to the east of the reserve. Buses also stop in Evie and in Dounby, seven kilometres west of the reserve.

Cottascarth and Rendall Moss (HY 369 195), with its wonderful turf-roofed stone hide is seven kilometres north of Finstown, off the A966. The minor road leading west at Norseman Village is signposted for RSPB Cottascarth. Turn right at Settisgarth and follow the signs for the reserve, passing through the farmyard to the car park beyond Lower Cottascarth Farm, and then follow the signs for the hide. Buses running between

Kirkwall and Tingwall stop at Norseman Village, three kilometres east of the reserve. Cycle Route 1 of the National Cycle Network runs through Norseman Village.

The island of Hoy can be reached from the island of Mainland by a passenger ferry that also carries bikes, from Stromness to Moaness on Hoy. From the Moaness pier, go straight up the hill and follow the signs for the Dwarfie Stane towards Rackwick. There is an information board after 2.5 kilometres on the Post Road footpath. A car ferry operates between Houton on Mainland Orkney to Lyness on Hoy. The RSPB has a presence during the breeding season to show people the eagles at a safe distance, along with other raptors, such as merlin (*Falco columbarius*), peregrine falcon and hen harriers.

If time allows, the upland summits on Hoy are worth exploring for their rare Arctic-Alpine flora and hard-won views of Scapa Flow and the rest of Orkney. There is no recognised path to the summits but as good a way as any is to continue north and east beyond the Old Man of Hoy, or else strike up the slopes either side of the glen to the west of Ward Hill.

The RSPB Trumland Nature Reserve on Rousay includes blanket bog, much of it cut-over for turves but still a good place to see red-throated divers and hen harriers. The island is reached by ferry from Tingwall on Mainland. From the pier, follow the road uphill to the junction at Trumland house and head right to a small bridge and RSPB sign on the left after 600 metres.

There are several Orkney-based wildlife tour operators who can provide detailed local knowledge, among which Orcadian Wildlife is well established and provides accommodation on South Ronaldsay.

Place Names

Birsay: Old Norse from *Byrgisey*, meaning 'fortress island' – an island accessible only by a narrow neck of land.

Hoy: Old Norse *Háey* meaning 'high island'.

Mainland: Old Norse *Meginland* with the same meaning.

Eday: possibly Old Norse *Eithey* meaning 'isthmus island'.

Cottascarth: Old Norse *Kottaskarð* meaning 'the hill gap with small houses'.

Rousay: Old Norse *Hrólfsey* meaning 'Rolf's Island'.

THE FLOWS OF CAITHNESS AND SUTHERLAND

Map: OS 1:50,000 Sheet Nos 9, 10, 11
Peatland Grid ref: NC 866 402
Access point: Forsinard (NC 890 425)

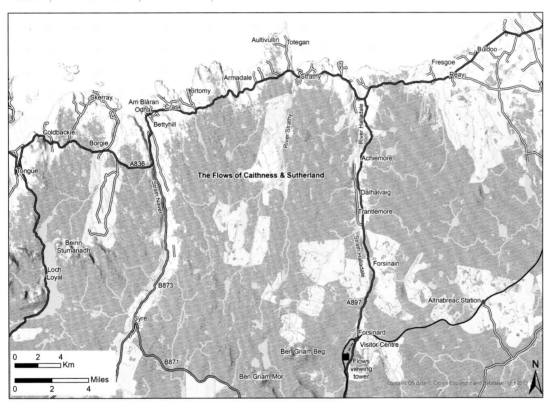

The Flows (pronounced to rhyme with 'cows') covers 400,000 hectares, largely in Sutherland and extending into Caithness. This is the biggest Atlantic blanket bog complex in Europe and with its important peatland bird populations it is the best example of the habitat in the world. The peat here can be more than five metres deep and stores more than three times the carbon contained in all Britain's forests. The Flow Country has been proposed as a UNESCO World Heritage Site, reflecting its 'outstanding universal value'. It truly deserves such an accolade.

Dubh Lochans, Forsinard Flows

The broad, flat, rolling expanse of blanket bog merges into wonderful scenic straths (valleys) lined by ancient woodlands, with mountain rivers prized by fisherman. The hills of Ben Griam Mor, and Ben Griam Beg pierce up through the low-lying blanket bog south-west of Forsinard, with Morvern inland of Berriedale offering incredible views of the peatlands stretching east to the plains of Caithness. Further west in Sutherland, the land becomes more mountainous and includes Ben Hope, Scotland's most northerly Munro (a mountain in Scotland over 3,000 feet or 914.4 metres in height).

The name Flows comes from the Old Norse *floi* meaning 'soft or marshy ground'. Timothy Pont's 16th-century maps show a distinctive area of 'moors and moss' south of Strathy Point between Strath Naver and Strath Halladale, with another to the north-east of Loch Loyal.

The RSPB Forsinard Flows reserve lies at the heart of this vast peatland and offers a perfect starting point for any visit, with a conveniently located information centre in the old railway station at Forsinard. The best way to appreciate the enormity of the peatland landscape is to take the boardwalk to the award-winning lookout tower offering panoramic views of bog and use the short trail of Caithness flagstone to meander through the small black pools, or *dubh lochans*. The Flows are prized for the huge numbers of breeding waders, such as dunlin, golden plover and greenshank, as well as both red-throated and black-throated divers (*Gavia arctica*). In such a large landscape these wary birds can be hard to spot, but they certainly can be heard making a variety of plaintive calls that echo across the peatlands in spring and summer. This is also a great place to see hen harrier, merlin and occasionally golden eagle (*Aquila chrysaetos*) as they search for prey among the bogs and heath.

Peatland plants are the building blocks of the Flows, with many different sphagnum mosses together with cotton-grasses, sedges and heathers forming patterns of reds, greens and yellows that shift as the seasons change. Look out for the ginger-biscuit-coloured hummocks of 'rusty bog-moss', a rare sphagnum that has its own Species Champion: a Scottish Environment Link initiative working with Members of the Scottish Parliament to adopt species and lend political support to their conservation. The glistening bog pools are criss-crossed by the stems and lobed leaves of bog bean (*Menyanthes trifoliata*), a favourite food plant of the red deer herds that traverse the bogs.

The Flows are among the strongholds in Britain for dwarf birch (*Betula nana*), a natural miniature bonsai tree that grows no more than a metre tall among the heather. Plantlife

Drawing: Black-throated diver

Forsinard flows Lookout Tower

has its 2,000-hectare Munsary Nature Reserve on the eastern edge of the Flows at Achavanich. Typical bog species can all found here, as well as the rare marsh saxifrage (*Saxifraga hirculus*) and the even rarer bog orchid (*Hammarbya paludosa*), a tiny yellow-green orchid, pollinated by midges, that is hard to spot among the mosses.

For thousands of years people have lived and worked mainly around the margins of the Flows, close to the coasts, along the river valleys and on the drier sloping hill ground. The bogs were places for grazing livestock and cutting peat for fuel on the more easily drained peatland edges and hillsides. A visit to Mary-Ann's Cottage at Dunnet Head shows what crofting life was like and how important peat was for peoples' survival. Today it is still possible to see stacked peat turves, but few communities now rely on this form of fuel.

Mechanised peat cutting in the 20th century extended further into the bogs, with various commercial enterprises exploiting the peatlands, including an experimental peat-fired power station at Altnabreac in the 1950s. Commercially cut sites such as seen at Forsinard supplied peat for the malting of barley by the numerous distilleries that can be found down the east coast of Caithness, although most of the malting is now done centrally in Inverness.

In the 20th century, much of the peatland was drained for agriculture, even on the areas of bog most challenging to reach. This government-funded attempt to make better grazing land for sheep ultimately failed, leaving a massive network of 'herringbone' and concentric drainage channels traversing much of the peatland, still visible in satellite images. Improvements in drainage techniques and more government funding brought commercial forestry planting to the Flows in the 1970s and 1980s. Tens of thousands of hectares of peatland were planted with mainly non-native conifers on land that had not naturally held trees except for a brief period 4,000 years ago and before that, since the peats started to form, following the end of the last Ice Age. After both the economic folly and environmental consequences were realised, the tree planting stopped, leaving a legacy of failing conifer plantations.

In 1988, conservation efforts saw the Flows designated in recognition of its international wildlife importance and in 1995 the RSPB established its largest UK nature reserve at Forsinard. Government grants and EU funding allowed a massive recovery operation, with conifers being cleared and drains blocked to re-wet the peatlands. Over the last two decades, a partnership approach has seen land managers and environment organisations agree mutually beneficial ways of looking after the peatlands and helping share knowledge about their importance.

In 2014, the Peatland Partnership began a multi-million-pound project managed by the RSPB to deliver a programme of restoration, education, visitor interpretation and science. A new research centre for peatlands was established at Forsinard, providing facilities to study the many important aspects of peatlands. Scientists from the Environmental Research Institute in Thurso, the James Hutton Institute and the University of Aberdeen all benefit from this new peatland centre of excellence.

One of the sad ironies is that research on peatlands has not kept pace with the threats of new developments. The advent of wind energy has seen large-scale construction of turbines and connecting roads on the peatlands, with limited understanding of their impact. One controversial scheme at Strathy South was granted permission by the Scottish Government in the face of objections from environmental bodies, including the government's own advisors NatureScot. The developers claimed that the windfarm development would include conifer tree removal and repair of previously damaged peatland. Conservation bodies argued that peatland restoration would be more effective and less risky to wildlife if it was done without the intrusion of a huge windfarm. Hopefully it is not beyond the skill of industry and good governance to plan development away from such a special peatland, enabling renewable energy and carbon-storing bogs without compromising both.

Travel Notes

The Inverness–Wick train stops at Forsinard, where the RSPB has a visitor information centre in the disused railway station building. The centre is also close to the A897, a single-track road with passing places that runs from Helmsdale in the south to Melvich in the north. The Dubh Lochan Trail and the Flows Lookout Tower are a short walk from the station.

Six kilometres north from the station on the A897 is the longer Forsinain trail – a marked, looped route starting on farmed land at NC 903 485 and extending out through bog and forest. A temporary diversion is in place while tree felling takes place for peatland restoration.

The Plantlife Munsary Nature Reserve is situated east of the A9 road at Achavanich and can be reached by a track starting from the car park overlooking Loch Stemster at ND 186 423. Well-signed posts lead five kilometres to the ruins of Munsary Cottage, with spectacular peatland views on the way. The return route is on the same track.

For a more challenging mountain experience of blanket bog in Sutherland, the Moine Path is a heritage route starting near the south end of Loch Hope and extending thirteen kilometres eastwards towards the derelict Kinloch House and the A838 at the Kyle of Tongue. From the minor road at the southern end of Loch Hope, a track at NC 458 507 leads north-east round the foot of Ben Hope. The vehicle track is obvious as it heads out, but it does narrow to a single track in the central section before widening out again on the descent over Allt Ach' an t-Strathain.

There is a Neolithic chambered cairn at NC 550 525 and shortly afterwards the track joins an estate road that heads northwards out of the strath to another ancient structure, the Dun Mhaigh Broch (NC 552 530). The minor road that runs around the southern end of the Kyle of Tongue leads onto the village of Tongue.

The public transport option involves a lengthy thirty-kilometre walk starting and finishing at Tongue. The Durness bus runs along the A838 with a stop at Hope. There is then a ten-kilometre walk south along the minor road on the east side of Loch Hope to the start of the Moine Path.

Place Names

Forsinard: from *Fors na h-Àirde* a hybrid of Old Norse (*forss*) for 'waterfall' and Gaelic for 'high point' (*àrd*), giving 'the waterfall of the high ground'.

Forsinain: anglicised Old Norse and Gaelic *Fors an Fhàine* 'the waterfall of the low ground'.

Achavanich: from Gaelic *Achadh a' Mhanaich*, 'the field of the monks'.

Mòine: Gaelic for 'moss' or 'boggy morass'.

Ben Griam Beg and Ben Griam Mor: from Gaelic, meaning 'small and large mountains of the sun'.

OUTER HEBRIDES

The remote, mountainous west coast of Scotland is flanked by a diverse group of islands split into two main groups: the Inner and Outer Hebrides. Alongside previous Norse influences there is a strong ancient Gaelic culture. The Gaelic language continues to thrive, particularly in the Outer Hebrides where the local authority council is officially known by the Gaelic name *Comhairle nan Eilean Siar*, meaning 'Western Isles Council'.

The major islands of the Outer Hebrides are: Lewis and Harris, a single island with two names; North Uist, Benbecula and South Uist, collectively known as The Uists; and Barra. Lewis, the northern part of the largest island, is comparatively flat and is dominated by blanket bog, with Harris to the south being more mountainous. The bedrock underlying these islands is mainly Lewisian gneiss, amongst the oldest rocks in the world, having been formed up to three billion years ago.

The Lewis peatlands are the second-largest blanket bog in UK, next in size only to the Flow Country. Blanket bog also covers more mountainous ground in the north of Harris and is frequently found across many of the Outer Hebrides islands.

The main land use in the Outer Hebrides is crofting, a form of tenant farming, mainly rearing sheep but increasingly supplemented by a wide range of business ventures, including hospitality and wildlife tourism. The dramatic scenery, world-class beaches and abundant wildlife along with a rich cultural heritage make these islands very popular tourist destinations.

Lewis and Harris are famed for the manufacture of tweed, a woven wool material with ancient origins. Traditionally hand-spun wool was coloured using dyes based on peatland plants and hand woven in crofters' homes. Today the initial processes are carried out in mills and then the tweed is hand-woven in homes across the islands before returning to the mills for finishing. Legislation requires that in the making of Harris tweed the whole process must take place in the Outer Hebrides.

Ferry links from Ullapool and Mallaig on the Scottish mainland and from Uig on Skye connect with the Outer Hebrides islands of Lewis and Harris and The Uists. Quiet, rural roads make cycling across the islands a relaxing yet thrilling experience.

Looking south to Harris near Lochganvich, Lewis

LEWIS PEATLANDS

Map: OS 1:50,000 Sheet Nos 8, 13, 14
Peatland Grid ref: NB 326 379
Access point: Loch na Muilne (NB 311 494), North Tolsta (NB 537 474)

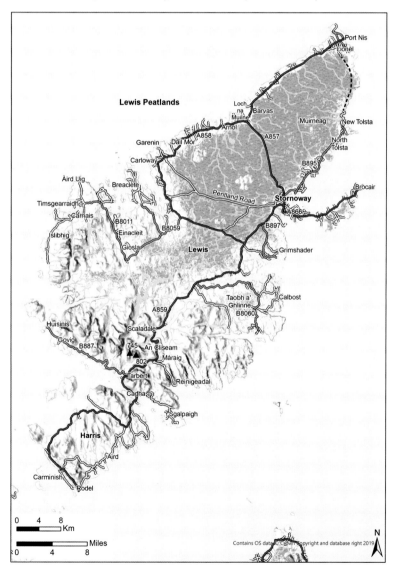

Eroding peat hags, Bheinn Bhragair, Lewis

The island of Lewis and Harris is often referred to as two separate areas although both are joined. In the north, Lewis with its subdued topography dominated by flat expanses of blanket bogs, contrasts with the rugged mountains of Harris.

Facing the Atlantic Ocean, the island of Lewis and Harris sits at Europe's westernmost edge. Sparsely populated, much of the land is uncultivated except for the unique 'machair' in coastal areas, where shallow peat is inundated by windblown calcium-rich shell sand to create fertile grassland, rich in wildlife. Place names on the island largely have Norse origins and the name Lewis itself is thought to relate to a 'songhouse', or it could be derived from the Gaelic *leogach* meaning 'marshy'. This ties in with a 2nd century reference to Lewis by Greco-Roman geographer Ptolemy, who named it *Limnu*, also meaning 'marshy'.

The Lewis peatlands occupy a large area of low-lying ground to the west and north of the main town of Stornoway. One of the largest blanket bogs in Europe, the Lewis peatlands have experienced less intrusion from forestry and windfarms than their counterparts in Caithness and Sutherland. The blanket bog is interspersed by numerous bog pools and freshwater lochs much favoured by black-throated and red-throated divers. One particularly characteristic feature of the Lewis peatlands is the widespread occurrence of the woolly fringe-moss (*Racomitrium lanuginosum*) that gives the tops of the peat hummocks a white, hairy appearance.

Peat cutting banks, Butt of Lewis

The island's main road runs along the western edge of the peatland with several crofting townships along the route. Gaelic is widely spoken among the communities here and there is a long tradition of grazing livestock on the peatlands as well as cutting the peat for fuel. The peatlands may appear vast and derelict, but the influence of people is still evident, particularly around the margins scarred by numerous peat-cutting banks. Although less relied on as a fuel nowadays, rows of stacked peat turves can still be seen.

Another feature scattered across the peatlands are the small buildings, some no more than huts used by the peat cutters. These shielings were originally built as summer residences when families, mainly women and children, would take their cattle, and more recently sheep, out to graze on the peatlands for the summer. Signs of human intervention go back thousands of years and studies of the soil layers beneath the peat suggest the island held large areas of native woodland that appear to have been burned, perhaps to provide grazing land for livestock.

Red deer are found on the Lewis peatlands. Archaeological evidence suggests they appeared with the arrival of Neolithic people. The island is too far from the mainland to be colonised naturally by deer, and DNA analysis shows the present-day population is more closely related to herds in northern Europe than they are to nearby mainland populations. They also share an ancestry with the ancient deer populations that were once found on Orkney and Shetland before becoming extinct there, which again suggests they were brought by the people from northern Europe who colonised all these islands.

One of the main towns, Barvas, gives its name to a large section of the peatlands knows as Barvas Moor. It was here in 2004 that a proposal for a 234-turbine windfarm on the peatlands caused an outcry among locals and from conservation bodies across the UK. Scottish ministers eventually refused the application on the basis that it would harm the internationally important bird populations of the area.

This was not the first attempt to bring major industry to the island. After the First World War, Lord Leverhulme, the soap baron, took ownership of the island and set about his plans to transform the peatlands into arable land providing crops of vegetables and soft fruit. The venture failed as he hadn't taken account of the huge peat depth to be stripped and the poor nature of the soil, let alone the harsh climate and its salt-laden winds.

Another attempt to exploit peat was in the 19th century by a previous owner of the island, Sir James Matheson, who built the Lewis Chemical Works near Stornoway in the 1850s. The factory burned peat and distilled the tar to make paraffin lamp oil, candles and sheep dip, before eventually closing down in 1874.

Today the islands are a haven for tourists who come to enjoy the peatlands for what they really are – a rich, spectacular wildlife experience in a wonderful island setting.

Peat stacks on Pentland Road, Lewis

Travel notes

Stornoway is the main port for ferries arriving from Ullapool on the mainland and has a small airport three kilometres to the east of the town. From Stornoway there are buses to all parts of Lewis, except on Sundays.

If cycling, the Pentland Road running from Stornoway to Carloway on the west of the island is a quiet single-track road providing excellent views over the Barvas Moor to the north. At Carloway there is an Iron Age broch, Dun Carloway, and a blackhouse village at Garenin with restored thatched crofters' houses. The Pentland Road, completed in 1912, is named after Lord Pentland, then Secretary of State for Scotland, who helped secure funds for its construction. There is an RSPB reserve Loch na Muilne (NB 311 494) about twelve kilometres along the A858 towards Barvas. The reserve lies just past the village of Arnol, which has a blackhouse museum. A short waymarked route gives views over peatland and coastal heath with boggy pools where red-necked phalarope breed.

There are relatively few hill paths on Lewis, as people traditionally just walked across the open moors. The most obvious hill on Barvas Moor is Muirneag (NB 479 489), whose summit is at 248 metres and offers great views of the peatlands but is only for the experienced walker as the terrain can be very boggy and there is no marked path.

The east and north of Barvas Moor can be experienced by way of an eighteen-kilometre waymarked trail between North Tolsta and Port of Ness, both connected by buses from Stornoway. The route follows the B895 road up the east coast to North Tolsta, and then passes the glorious sandy beaches of Tràigh Mhòr and Garry Beach before crossing the 'Bridge to Nowhere' (NB 531 501). This was originally planned by Lord Leverhulme to be the start of a road to Ness, but the project was never completed. From the end of the road, follow the unfenced track and waymarker posts which lead onto tough-going, boggy ground before reaching a gravel track and the road at Sgiogarstaigh. Continue north along the road to the bus stop at Lionel, by Port of Ness, to return to Stornoway.

A long-distance walking and cycling route, the Hebridean Way, follows part of the Pentland Road before heading south 250 kilometres through Harris and The Uists to the south of Barra. The route is divided into twelve main sections and includes splendid moorland sections and hilltops with views over the peatlands, including the area to the west of Loch nam Madadh (Lochmaddy), with some of the best examples of blanket bog on The Uists.

Place Names

Carloway: Gaelic *Càrlabhagh*, originally Viking meaning 'Karl's Bay'.

Stornoway: Gaelic *Steòrnabhagh*, from Norse *Sjornavagr* meaning 'steering bay'.

Muirneag: Gaelic diminutive of *muirn*, 'cheerfulness, joy'.

Garry: Gaelic *geàrraidh*, meaning 'an enclosed grazing between the arable land and the open moor'.

Tràigh Mhòr: Gaelic for 'big beach'.

Sgiogarstaigh: Old Norse, 'Skeggi's place'.

Loch na Muilne: Gaelic 'loch of the mill'.

Loch nam Madadh: Loch of the hounds (*madadh* in Gaelic), from the rocks at the entrance to the loch, said to resemble dogs' heads.

GRAMPIAN MOUNTAINS

The Grampian mountain range extends across the Scottish mainland from south-west to north-east between the Highland Boundary Fault and the Great Glen. Opinion differs as to whether it is a single range or a number of distinct ranges. Even the origin of its name is confusing, thought to be from the Roman description of the Battle of Mons 'Graupius' (*Graupius* having no known meaning), before being adopted in the Middle Ages as Grampians by the philosopher and historian Hector Boece. The range contains Britain's two highest mountains, Ben Nevis in the west and Ben Macdui in the east.

This sparsely populated and remote region provides a dramatic and diverse landscape in which some of the best examples of all three major peatland types – fen, raised bog and blanket bog – can be found.

Below the Cairngorms is the great expanse of the Insh Marshes, lying in the River Spey floodplain and forming the largest river fen in Britain. Flanking the Cairngorms on the western side of Strathspey the Monadhliath Mountains support large areas of blanket bog, much of which is eroded with extensive gullying. Conservation work is underway to tackle the drainage and red deer grazing and trampling which had given rise to the damage. The Cairngorms National Park Authority is helping restore damaged peatland habitats across the national park.

In the west of the region is the Great Moss, Moine Mhòr, a large raised bog at sea level in a region that was a focal point for Scotland's ancient tribes. Between these two magnificent peatland areas is the equally impressive blanket bog of Rannoch Moor, guarding the entrance to Glencoe and intimately bound to the turbulent history of the Highland clans who lived there.

Despite their remoteness, all three peatlands have had major interventions and attempts to drain them for agriculture, industrial activity and forestry. In part due to their great size, they have retained much of their peatland character and provide a great wildlife spectacle as well as an emotive connection to the lives of people stretching back thousands of years.

Travel in these remote areas, particularly in the west, is challenging but there are bus and rail connections to all three sites. Insh Marshes contains an RSPB nature reserve and there are exhibits at Rannoch Station and Moine Mhòr explaining the rich history of these sites.

Entrance to Tileworks Trail, Moine Mhòr

INSH MARSHES

Map: OS 1:50,000 Sheet No 35
Peatland Grid ref: NH 780 013
Access point: Insh Marshes Reserve (NN 775 998)

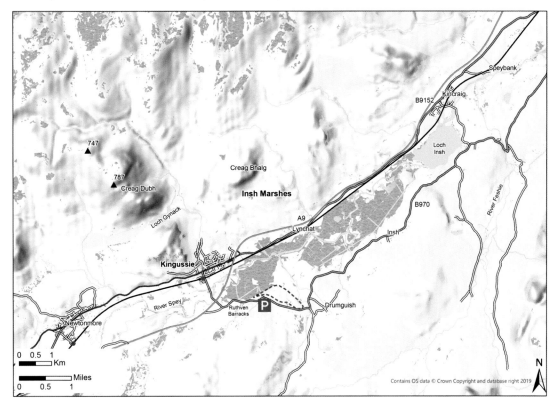

Set in the foothills of the Cairngorm and Monadhliath mountain ranges, the impressive Insh Marshes lies in a huge floodplain of the River Spey as it winds its way between Kingussie and Kincraig in the Scottish Highlands. This is one of the best remaining examples of a river fen system in the UK, with the powerful waters of the River Spey bursting their banks and inundating the surrounding area several times each year. By providing a natural water storage system during times of spate, Insh Marshes helps reduce the risk of flooding to downstream villages. The combination of fen habitat and open flood water covering over a thousand hectares makes this one of Europe's most important wetlands and it is managed by the RSPB as a nature reserve. Large numbers of breeding waders, such as redshank (*Tringa totanus*), snipe and curlew, and migrating waterfowl in autumn

Transition mire at Insh Marshes

and winter including up to two hundred whooper swans (*Cygnus cygnus*) from Iceland, provide an all-year-round wildlife spectacle.

Insh Marshes is a type of peatland called a 'transition mire' that includes elements of both fen and bog. The soils here are generally nutrient poor and acidic giving rise to 'poor' fen dominated by short *Carex* sedges and supporting sphagnum bog in parts. This is one of only two sites in the UK where the rare string sedge (*Carex chordorrhiza*) can be found, along with a remarkable five hundred other plant species. Ancient stands of aspen trees (*Populus tremula*) are some of the best found in Scotland and provide a beautiful shimmering feature that turns brilliant yellow then red in autumn.

For centuries, attempts to drain the marshes to provide improved agricultural land simply led to poor-quality grazing that was eventually abandoned, leaving the damaged site at risk of being overgrown by willow scrub. Conservation work, including cutting down the willow and working with local farmers to allow low-density grazing by cattle, sheep and ponies, is allowing the fen vegetation and herb-rich meadows to recover.

In 2018, plans were announced for upgrading the A9 trunk road that passes through the Insh Marshes Nature Reserve, which could damage important breeding wader habitat. The RSPB proposed alternative options to allow for road improvements while conserving the wildlife.

Overlooking the marshes are the distinctive ruins of Ruthven Barracks, situated on a natural mound with a dominant position at the head of the Spey Valley. Built following the Jacobite rising of 1715, it was designed to house around 120 of the government's infantry. Overlooking the confluence of three military roads, it was an ideal location to control the surrounding area. Later, during the 1745 rising, the Jacobites took the barracks and set them ablaze. After their defeat at the Battle of Culloden, the surviving Jacobite forces regrouped here awaiting instructions from their leader Bonnie Prince Charlie, only to be given the heartbreaking news that they were to disperse, with the prince then departing Scotland's shores.

Travel Notes

The nearest train station and bus stop are at Kingussie, around 2.5 kilometres away. The RSPB nature reserve entrance is one kilometre to the east of the Ruthven Barracks on the B970. The reserve is on National Cycle Route 7.

There are two hides on the reserve and two trails giving wonderful panoramic views of the marshes and surrounding countryside as well as going through some of the adjacent ancient woodlands.

The villages of Kingussie and nearby Newtonmore provide accommodation shops and toilet facilities.

Place names

Monadhliath: Gaelic *monadh liath* meaning 'grey-blue mountain range'.

Insh: Gaelic *innis*, meaning 'island' or 'meadow'.

Spey: from a Brythonic word *yspyddad*, meaning 'hawthorn river', or derived from the pre-Celtic *squeas* meaning 'gush'.

Ruthven: from Gaelic *ruadh* 'red', as in *Taigh-feachd an Ruadhainn* meaning 'red barracks'.

Kingussie: Gaelic *Ceann a' Ghiuthsaich*, meaning 'head of the pine forest'.

RANNOCH MOOR

Map: OS 1:50,000 Sheet Nos 41, 42
Peatland Grid ref: NN 373 523
Access point: Bridge of Orchy (NN 297 396), Rannoch Station (NN 422 578)

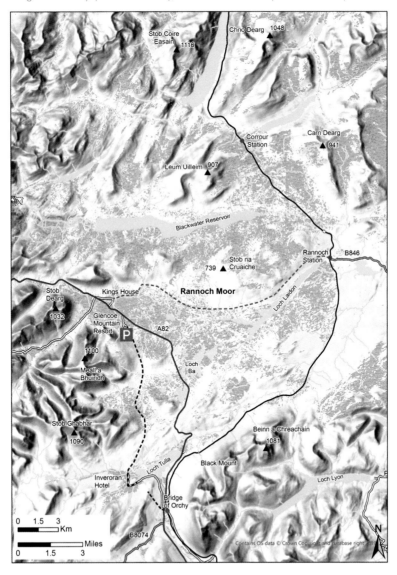

The vast expanse of Rannoch Moor is situated on a high-level plateau three hundred metres above sea level and covering over 130,000 hectares between Loch Rannoch, Glencoe

Rannoch Moor looking over Corrour Station

and Bridge of Orchy. The moor is mostly blanket bog but is notable for its varied ground profile carved by successive periods of glacial activity that has formed a mix of rocky knolls and valleys supporting bog, fen and heath. A large remnant of ancient pine and birch wood, which once surrounded the peatland, can be seen at the Black Wood of Rannoch to the south of Loch Rannoch.

This is the best remaining peatland site in Britain to find the rare Rannoch-rush (*Scheuchzeria palustris*). The plant favours wet peat and occupies the edges of mossy pools and small fens within the bog. Once far more widespread, it has been entirely lost to other parts of Britain and Ireland since the end of the 19th century through drainage and peat cutting.

Rannoch Moor occupies a commanding position to the east of Glencoe and is surrounded by the mountains of Ben Alder, Carn Mairg and Blackmount. Its dramatic setting and association with tales of the Highland clans, including the massacre of the MacDonalds of Glencoe by government forces, has made this one of the most famous and intriguing of Scotland's peatlands. Chilling mists can rapidly descend in this lonely landscape and the sense of eeriness has been unsettling for many writers who travelled here. Robert Louis Stevenson's book *Kidnapped*, published in 1866 but set in the period just after the Jacobite rebellion, refers to the moor as a 'country lying as waste as the sea; only the moorfowl and pewees crying upon it… A wearier desert man never saw'.

The travel writer, the Reverend John Lettice, in his *Letters on a Tour Through Various Parts of Scotland, in the Year 1792*, described Rannoch Moor as an 'immense vacuity with nothing in it to contemplate'. Our contemporary travel writer Cameron McNeish, on the other hand, describes the same scene in his 1999 book *Scotland's 100 Best Walks* as 'immensely appealing, an empty quarter where the spirit can soar in unfettered abandon', and goes so far as to suggest that such a moor can share the same attributes as our highest mountains.

The remoteness of Rannoch Moor has not saved it from the interventions of human economic activity. In the 18th century, shortly after the Jacobite rebellions, when the lands were forfeit to the Crown, the appointed estate factor James Small set about a largely unsuccessful attempt to 'improve' the moorland through employing the military to create a large network of drains. Some of these 'soldiers' trenches' can still be found today. It is interesting that Stevenson's *Kidnapped* description also refers to blackened, burned areas and haggs – columns of exposed eroding peat – on the moor, suggesting that drainage and combinations of grazing and fire management had taken place.

Drawing: Golden eagle (immature)

In 1889, work began on constructing a railway line across Rannoch Moor to connect Glasgow and Fort William, later extended to Mallaig to complete the West Highland Line. Laying the tracks to accommodate heavy trains on deep peat was no easy engineering task. Thousands of tonnes of ash and rock were deposited on the peat, which gave limited support, and it took many years of rebuilding before the railway line was stabilised using bundles of brashwood (the thin lower tree branches removed during forest thinning). Even now, passing trains create a vibration that can be felt rippling through the peat by nearby walkers.

Travel Notes

The southern part of the moor can be reached from the Bridge of Orchy railway station by following the West Highland Way, with a steep ascent over Mam Carraigh and down to Inveroran Hotel by Loch Tulla. A few kilometres after crossing the Victoria Bridge, the blanket bog and small lochans across Rannoch Moor become visible. The route continues to the Glencoe Mountain Resort beside the A82, where a bus can be taken back to Bridge of Orchy.

Rannoch Station has a tearoom and a small visitor centre, opened by David Bellamy in 2005, that holds displays explaining the history of Rannoch Moor and describing the important peatland wildlife. For excellent views of the peatland, an ancient drove route leads from Rannoch Station twenty kilometres west to the Kingshouse Hotel, originally built in the 19th century as part of the government's improved access after the end of the Jacobite rising.

From Rannoch Station car park, take the small road west over the railway and through a forest plantation on the north side of Loch Laidon. After a gradual climb, the route becomes boggy and contours the hillside following a line of electricity pylons before reaching a track to the Black Corries Lodge and onto the Kingshouse Hotel. A two-kilometre walk eastwards along the A82 reaches the bus stop at the Glencoe chairlift road end, for onward travel to the railway station at Bridge of Orchy and Glasgow. There is a youth hostel at Loch Ossian as well as bed and breakfast accommodation and a restaurant at the 'Signal Box' beside Corrour Station, the next stop north of Rannoch Station.

Place names

Rannoch: Gaelic *raineach* 'bracken/fern'.

Carn Mairg: Gaelic 'hill of sorrow' or 'hill of the dead'.

Mam Carraigh: Gaelic 'round hill of the rocky cliff'.

MOINE MHÒR

Map: OS 1:50,000 Sheet No 55
Peatland Grid ref: NR 815 925 South Moss, NR 828 950 North Moss
Access point: Tileworks Trail (NR 825 958), Crinan Canal (NR 804 923)

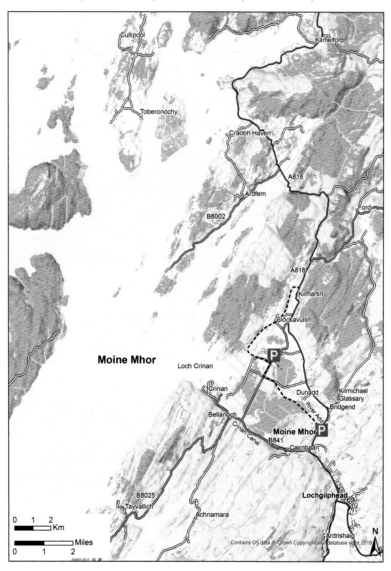

Set in the ancient heart of Scotland at the mouth of Kilmartin Glen in Argyll, the Great Moss, Moine Mhòr, is a magnificent example of estuarine raised bog habitat. The peatland

The Great Moss, Moine Mhòr

Woodland sculpture, Tileworks Trail, Moine Mhòr

is also associated with one of the greatest concentrations of Neolithic and Bronze Age remains in Scotland.

The Moine Mhòr sits on a low-lying flat plain surrounding the River Add. In what was once a large freshwater loch, peat began to form around 5,500 years ago as the climate became milder and wetter. Originally covering over 1,600 hectares, the peatland consisted of three distinct domes, South Moss, West Moss and North Moss. In the 19th century, the area around West Moss in particular was lost through drainage for agriculture and peat cutting to fuel a tile works that supplied pipes and bricks for the development of Kilmartin village. Later commercial forestry plantings on the peat caused further damage. Fortunately, North Moss was largely spared and is one of the few uncut original raised bog peat surfaces in Scotland, with peat as deep as four metres, having built up millimetre by millimetre over the millennia.

In 1987, much of the remaining peatland was designated as a National Nature Reserve and has been managed by NatureScot, which has carried out extensive works blocking old drains and helping the peatland vegetation recover. Plantation conifers on the peat have been felled, which helps the bog wildlife and retains the massive carbon store held within the deep peat.

Small pockets of the original lagg fen habitat with reeds and alder scrub occur at South Moss, and further east at Coille Mhòr near Barnakill there are sizeable remnants of the ancient oak woods that once grew across much of this landscape.

Around Moine Mhòr and into the broad valley of Kilmartin Glen are numerous Neolithic stone circles, rock carvings and burial chambers. By the Bronze Age this was a major area for people to travel to, with all the various monuments forming a connected experience. This was a hub for trade, with copper from Ireland and tin from Cornwall brought together to supply the demand for bronze in other parts of Scotland.

On a large rocky outcrop to the east of the peatland is the site of Dunadd, a hill fort that during the period between 500 and 900 CE was a capital of the Kingdom of Dalriada and a place where kings were anointed. On a stone platform below Dunadd summit, a deep footprint was carved into the rock for the ritual placing of the king's foot during his coronation to symbolise being joined to the land he ruled. From this vantage point, looking out over Moine Mhòr up the huge glen surrounded by high mountains and out over the saltmarsh to the sea, which would have been the main transport link for these ancient people, it is easy to get a sense of why this awe-inspiring place held such importance.

Travel notes

Buses run from Lochgiphead to Kilmartin to the north of Moine Mhòr and Crinan in the west. A fifteen-kilometre trail along paths and minor roads lead from the museum and café at Kilmartin around the peatland and along the Crinan canal to the village of Crinan. This route includes many of the main ancient monuments and passes Dunadd for spectacular views over the moss.

NatureScot has built a six-hundred-metre all-abilities Tileworks Trail that starts at a car park (NR 825 958) on the B8025. The trail leads through ancient Atlantic oak wood where the trees are festooned with mosses and lichens, and heads out onto a boardwalk across part of the North Moss before returning back to the car park.

Accommodation and food are available at Kilmartin and Crinan. It is well worth a visit to Crinan Ferry, reached by a minor road that crosses the canal at the mouth of the River Add. The Winterton bed and breakfast here provides a wonderful view across Moine Mhòr.

Place Names

Cnoc na Moine: Gaelic 'hillock/knoll of the moss'.

Coille Mhòr: Gaelic 'great wood'.

Barnakill: from Gaelic bàrr na coille meaning 'top of the wood'.

Dunadd: from Gaelic dùn 'fort' on the River Add, from 'fhada' meaning 'long'.

CENTRAL LOWLANDS

This low-lying region extending across southern Scotland is a broad valley sandwiched between the Highland Boundary Fault and the Southern Uplands Fault. Rich in coal and iron-bearing rocks and good agricultural land, the area supports the cities of Edinburgh, Glasgow, Dundee and Stirling, and holds over eighty per cent of the Scottish population. The undulating plains of the Central Lowlands contain the major river valleys of the Forth, Clyde and lower reaches of the Tay. Poorly drained soils are a key feature of the landscape and have allowed peaty soils to develop with raised bogs in the lowest ground and blanket bog in the hills. Many of the raised bogs are associated with coal deposits which are in effect just very old peat, formed millions of years ago in the wet, forested soils.

The once extensive wetlands of the Central Belt have been greatly reduced over the centuries by major drainage works to create agricultural grazing and arable land. In the 20th century, large areas were stripped of peat for opencast coal mining and much peatland was planted with conifer forests. Commercial mining of peat, primarily for horticulture, has taken place on numerous raised bogs in the region and continues at several sites to this day.

Being such a heavily populated and industrial area, the remaining peatlands include some examples very close to the major cities and have therefore become popular places for access to the countryside and enjoyment of the open air and wildlife. In recognition of this importance there are numerous peatland nature reserves. Local communities have shown great interest in their peatlands and there are several sites where groups have formed to take an active role in managing and conserving the peatlands, with participants describing the activities as a great healthy social event in an invigorating outdoor natural gym.

With so many excellent peatlands worth visiting in the Central Belt, it is difficult to select only a few highlights. This chapter presents the larger and more accessible peatlands to provide an initial experience of the region's peatland wealth.

Not detailed in this book but worth a mention are the peatlands included in the East Ayrshire Coalfield Environment Initiative, a partnership between the local authority and conservation bodies aimed at restoring peatlands with support from local communities. Two sites being restored as part of this project are Dalmellington Moss, a raised bog on the floodplain of the River Doon, managed by the Scottish Wildlife Trust, and Airds Moss, a blanket bog in the Muirkirk uplands, owned by the RSPB.

Travel in the Central Belt is made easy by a good network of train and bus routes between Glasgow and Edinburgh as well as quiet roads for cycling.

Dragonfly sculpture, Blawhorn Moss

FLANDERS MOSS

Map: OS 1:50,000 Sheet No 57
Peatland Grid ref: NS 630 985
Access point: Flanders Moss car park (NS 647 978)

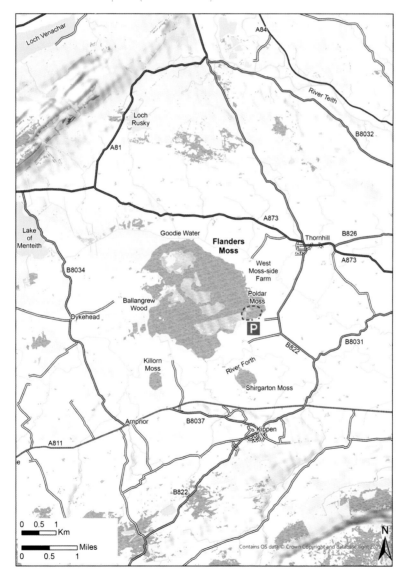

Flanders Moss is a National Nature Reserve and lies in the floodplain of the Forth estuary, fifteen kilometres west of Stirling. There is much speculation about the derivation of the

Viewing tower at Flanders Moss

name Flanders. It is unlikely that there is any association with Flemish immigration to Scotland in the 12th and 13th centuries as the name existed well before this date, nor that it was bestowed by 17th-century soldiers returning from European wars where similar peatlands occur in the Flanders region of northern Belgium. Although there is no definitive answer it appears that the name may have Gaelic roots, with both Pol*dar* and Flan*ders* having a connection with 'black', *dubh*.

At over 860 hectares in area, Flanders is the largest of a cluster of bogs in the Carse of Stirling. The word carse (the modern form of Old Scots *kerse*) refers to low and fertile land alongside a river. Although greatly reduced in area, the remaining peat dome (up to seven metres deep) at Flanders is the largest raised bog in Britain and retains most of its original peat profile.

On the west of the site at Ballangrew Wood there is a good example of the original lagg fen habitat, a feature that has been lost from so many raised bogs through agricultural activity. On the eastern edge, at West Moss-side Farm, a lagg fen creation scheme has been completed to restore this important transition habitat between farmland and bog. This is designed to benefit both the bog and the farm, and to improve habitat for farmland waders.

Looking north across Flanders moss from the viewing tower

In addition to the spring and summer displays of dragonflies, curlews and snipe, Flanders Moss is an important winter roost for hen harriers and there are thousands of pink-footed geese (Anser brachyrhynchus) at this time of year on the bog and surrounding fields. The site also holds important rare moss species Sphagnum austinii and Sphagnum majus that are found in hummocks in the least damaged parts of the site.

The peat has preserved a wide array of cultural artefacts including Neolithic trackways, Bronze Age swords and a bronze cauldron. There is a good display in the Stirling Smith Art Gallery and Museum which tells the story. The deep peat has also preserved the underlying geological information that provides some of the best evidence of sea level changes since the last glacial period, stretching back 12,000 years. Much of the surrounding carse is known to have been inundated by rising sea levels around 9,000 years ago, with preserved whale skeletons from this period found under the peat. Sea levels fell 8,000 years ago and peat began forming to create the present day raised bogs. Within the peat there is a layer of sand dated to 7,000 years ago, when a tsunami originating in Norway hit Scotland's east coast, sending material onto the bog.

For thousands of years the bog remained relatively unchanged apart from localised peat cutting for fuel at the bog margins. In the 18th century the peatland began to be cleared for agriculture. Landowners such as Lord Kames of Blair Drummond enthusiastically deployed a cheap labour force to strip away the peat to get to the mineral-rich marine clay soils below. Highlanders dislocated by the Clearances, and the aftermath of the Jacobite uprisings, were offered work and rent-free leases on strips of bog called 'pendicles' if they cleared the peat. The 'moss lairds', as they were ironically named, created several thousand hectares of agricultural land by cutting and dumping the peat into water-filled ditches that flowed into the River Forth and out to sea. Some of the peat works caused problems for downstream shell fisheries and other riverside communities and there were complaints of hundreds of thousands of tons of peat sediment and moss silting up the River Forth at Stirling, preventing access to the town by sea-going ships.

Fortunately, the deepest peat on Flanders Moss proved too difficult to remove and it was largely saved, having been reduced to sixty per cent of its original size by the time that large-scale peat stripping ended in the 19th century with reduced agricultural land value and higher labour costs. The following period of agriculture recession meant other uses were found for the peatland, and in the 19th century Flanders Moss became a grouse moor with drainage and regular burning to encourage heather. In 1878 a huge fire blazed over 900 hectares and took three days to put out. The grouse moor lapsed in the mid-1970s.

Forestry planting of commercial conifers was carried out on the peat in phases throughout the 20th century. The problem for the peatland was not only the associated drainage and loss of habitat below the trees but also the expansion of birch scrub woodland on the damaged and relatively dry bog surface. Peat cores and old maps show the

Flanders Moss trail

bog had been treeless for most of its history, until the start of the 20th century. The few trees that did occur are marked on old maps, presumably because they were important landmarks in this otherwise featureless landscape. Associated with the early woodland planting, rhododendron (*Rhododendron ponticum*) was introduced and has spread onto the open bog as well as a related plant, Labrador tea (*Rhododendron groenlandicum*), which normally grows in the high arctic and is believed to have been introduced rather than occurring naturally.

In the early 1980s, despite being a designated Site of Special Scientific Interest, a hundred-hectare section of Flanders Moss was cleared and levelled for commercial milling to provide gardening peat. This sparked a great deal of controversy and was one of the focal points for the conservation campaign against such destruction. In 1995 the planning permission for peat extraction was bought out by NatureScot and there began a major period of restoration management, including removing plantation trees, clearing birch and blocking drainage ditches. Low-density sheep grazing was introduced to suppress the growth of birch during the bog's recovery, before it eventually becomes too wet for trees. On the east side, West Moss-side Farm has introduced traditional Shetland cattle in the summer months to graze and browse the regenerating birch on the Poldar Moss section of Flanders Moss.

In the first decade of the 21st century, access to the site for visitors was improved through construction of a boardwalk and a viewing tower on the Poldar Moss. The area is recovering well from being drained as preparation for peat extraction, and in summer resembles a white sea of fluffy cotton-grass heads.

Travel notes

The nearest train station is in Stirling. Buses stop at Thornhill, about five kilometres from the entrance to the site, which is signposted from the B822 with a track leading west to the car park. The Glasgow to Callander section of National Cycle Route 7 passes near Flanders Moss. Leave the route at Braeval (south of Aberfoyle) and follow the A81 for seven kilometres and then turn onto the A873 heading towards Thornhill before turning south onto the B822 and the reserve car park where bikes can be left at a cycle rack.

From the car park, leave by a wooden bridge and join a path leading to a 900-metre boardwalk that loops around part of the moss. A seven-metre-high wooden viewing tower provides excellent views over the site.

For a longer day's experience, a 32-kilometre 'Tour de Carse' follows some of the minor roads in the area, taking in some of the key features in the surroundings that have shaped this historic landscape. There is food and accommodation available at Thornhill and Kippen. At West Moss-side Farm, Trossachs Yurts provides a luxury experience where conservation management work can be seen.

Place Names

Poldar Moss: from Gaelic *poll dubh*, 'black pow', a small slow-moving ditch or stream.

Kippen: from the Gaelic *ceap*, meaning 'cape' or 'promontory', or from *ciopan* meaning 'little stump', referring to the roots of trees which are a feature of the moss.

CLYDE-FORTH MOSSES

Map OS 1:50,000 Sheet Nos 64, 65, 72
Peatland Grid ref: Lenzie Moss NS 646 720, Cranley Moss NS 931 479, Blawhorn Moss NS 885 682, Red Moss of Balerno NT 164 636
Access point: Lenzie Moss (NS 639 511), Braehead (NS 955 507), Blawhorn Moss (NS 878 676), Pentlands regional car park (NT 166 638)

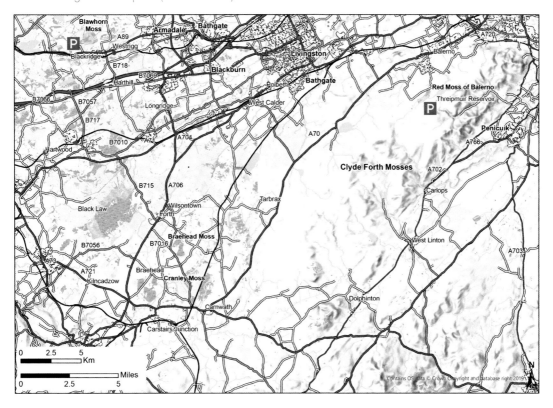

There are several raised bogs on low-lying land between the Firth of Forth and the River Clyde in a corridor that holds the highest human population density in Scotland. These mosses are popular as places for local people and visitors alike to enjoy open natural spaces.

In the west, around the city of Glasgow, Lenzie Moss and Langlands Moss are raised bogs managed by local community groups. In South Lanarkshire to the north of Carstairs, Cranley Moss and nearby Braehead Moss are typical of the domed, lens-shaped peatlands set in an agricultural landscape. North of the M8 motorway and west of Armadale is

Blawhorn Moss, archegonium sculpture, by Jim Whitson 'The Blazing Blacksmith'

Boardwalk Trail, Blawhorn Moss

Blawhorn Moss, a National Nature Reserve. Nestled at the foot of the Pentland Hills on the south-west edge of Edinburgh is the Red Moss of Balerno, managed by the Scottish Wildlife Trust.

The origin of these peatlands usually starts with depressions known as 'kettle holes', formed by ice during the last Ice Age and filled with water as the ice melted around 10,000 years ago. These wetlands then developed fen vegetation which succumbed to the growth of sphagnum mosses and the build-up of peat to form the distinctive visible domes that can be several metres deep.

Langlands Moss on the southern fringe of East Kilbride, a major industrial town, was recognised as an important open natural space and declared a local nature reserve in 1996. Much of the peatland had previously been drained and planted with commercial forestry. Conservation work in the mid-1990s saw more than 10,000 trees being removed by helicopter as well as the blocking of numerous drains across the site. The Friends of Langlands Moss, formally constituted in 2006, are a group of local people committed to conserving and restoring the site and protecting it against damaging development.

Lenzie Moss similarly occupies a major built-up area in the north-east of Glasgow. The peatland is divided by the main Glasgow to Edinburgh railway line. The site has a long history of peat cutting stretching as far back as the 13th century when peat provided fuel for the monks at Cambuskenneth Abbey in Stirlingshire. Commercial peat extraction for horticulture began after the Second World War and continued to the 1960s, by which time most of the original peat dome had been removed. The Friends of Lenzie Moss was founded in 1985 to help conserve the site, fending off a major housing development in 1990. Now a local nature reserve, work to restore the site includes the removal of birch tree seedlings and the installation of dams to help raise water tables.

Cranley Moss and Braehead Moss belong to a cluster of raised bogs centred around the villages of Forth, Carnwath and Carstairs in South Lanarkshire. Designated as internationally important raised bogs, they are part of Butterfly Conservation's 'Lanarkshire's Large Heaths and Mosses Project', aimed at restoring and conserving the sites. The peatlands in this area have experienced a wide range of pressures from agricultural drainage, commercial forestry, windfarm development and peat mining for horticulture.

The destructive impact of peat mining on peatland habitat can be seen at nearby Ryeflat Moss and Woodend Moss, where commercial extraction of the peat by a process known as 'milling' still takes place. North-west of the village of Forth is Black Law windfarm, one of the largest onshore windfarms in the UK. ScottishPower Renewables built the windfarm on a large area of blanket bog which had historically been drained and planted with conifers. Construction of the windfarm involved felling the trees and restoring the bog habitat and is now being monitored with the support of conservation bodies to study how the wildlife recovers.

Blawhorn Moss is one of the largest and least disturbed of the raised bogs in the Scottish Lowlands. Part of the site also merges into blanket bog where the peat is shallower and extends into the surrounding hills. The entire peat surface was drained in the 1940s but fortunately was never cut. Much of the site lies on the coal beds that led to the industrialisation of this region, and peatland on the west side of Blawhorn Moss came under the ownership of the National Coal Board, which managed a neighbouring opencast coal mine. Fortunately, the moss was spared from any peat removal and instead ownership was transferred to the nation, for the nominal sum of £10, and has been managed by NatureScot since 1980.

Conservation works in the late 20th and early 21st centuries focused on blocking ditches and reducing livestock grazing pressure. Fires thought to be started by vandalism have been a problem for this site but hopefully the risks of damage are lessened by the work to re-wet the peat. The community of nearby Blackridge village have been involved in developing access routes across the peatland as well as installing bog-inspired artworks produced by Jim Whitson, known as the 'Blazing Blacksmith'. Despite its proximity to a high-population area, Blawhorn Moss offers a peaceful haven with breeding red grouse, curlews, snipe and, in winter, hen harrier and short-eared owl (*Asio flammeus*).

Cranley Moss, South Lanarkshire; a lowland raised mire where surrounding natural carr woodland has been lost through conversion to agriculture.

The Red Moss of Balerno is a small twenty-five hectare raised bog in the Pentland Hills Regional Park. It was once a much larger bog but centuries of peat cutting and agricultural drainage around the edges have reduced it to the present size. What remains is a relatively undisturbed deep dome of peat largely protected from past damage as this was common grazing land. In the 18th century, pits were dug around the edges of the bog for the retting of flax. These were essentially wet pits where the fibres of the flax plant were broken down by acidic fermentation to prepare them for making linen. The industry ended in the 1780s with the arrival of cotton. An aqueduct constructed in the 19th century runs under part of the bog and was constructed to supply Edinburgh with water from springs in the Pentlands. Now abandoned, it is thought the collapsed channel may be responsible for partly draining the moss on its southern edge.

Travel Notes

This group of mosses between Glasgow and Edinburgh is easily accessed by public transport and provides a great opportunity for a cycling weekend excursion.

The nearest railway station to Langlands Moss is East Kilbride. Langlands Moss main entrance on Langlands Drive, near the junction with Hurlawcrook Road (NS 639 511), is signposted from Torrance roundabout on the A726 Strathaven Road in East Kilbride, just south of the Calderglen Country Park entrance. A footpath, the South Bridge Nature Trail, leads from the country park visitor centre alongside the west bank of the Rotten Calder Water to the entrance. There is a boardwalk on the bog and routes through the surrounding woodland.

Lenzie Moss is best accessed from the Lenzie railway station car park, where signs lead to the bog half a kilometre away. There is a two-kilometre circular signposted route around the site.

Braehead Moss is reached from Carstairs Junction railway station. There are several raised bogs in the vicinity which are well worth including on a circular cycle trip from the station towards the village of Forth. From Carstairs Junction head west towards Carstairs then turn right on the A70, and at the next junction turn left on the A721 heading west. At the next roundabout take the third exit onto the A706 and then the first right onto a minor road heading north-east. Cranley Moss can be seen on the right-hand side.

Follow the road to the village of Braehead on the B7016. The entrance to the Braehead Moss is between the school and the Last Shift Inn on the main road beside a school sign (NS 955 507). A track known as the Forest Walk leads to a short boardwalk with views over the bog. Head north-west out of the village on the B7016 and about 300 metres from the edge of the village turn left onto the Bog Road which joins the A706 and then turn right towards Forth.

Once in the village, take the turning left onto the B715 Climpy Road and after two kilometres the public entrance to the Black Law windfarm is on the right. Returning to Forth, take the A706 north-east out of the village and then right onto the B7016, where it is well worth a visit to the Wilsontown former iron works.

Continue south on the B7016 towards Braehead and two kilometres afterwards pass Scott's peat extraction site at Woodend Farm. Another kilometre down the road is Carnwath Moss, where the Forestry Commission Scotland has cleared trees to restore the peatland. Before reaching the junction with the A70, turn right to pass the White Loch and then a few hundred metres on the right is the Ryeflat Moss, where commercial peat extraction can be seen from the road. At the next junction turn left onto Ryeflat Road leading back to the A70 and return to Carstairs and the railway station at Carstairs Junction.

Blawhorn Moss is approximately ten kilometres from the nearest railway station at Bathgate. There is a bus stop in Blackridge. The reserve is reached by heading west from the village on the A89. After five hundred metres take the signed access road on the right that leads to a parking area. Marked routes lead to the bog and a small circular boardwalk as well as routes around the perimeter of the site. A connecting path leads from the bog back to Blackridge village. The stable block of the Innhouse at Blackridge has a display telling the story of the area.

For the Red Moss of Balerno, Lothian Buses number 44 stops at the Cockburn Crescent junction on Mansfield Road, approximately two kilometres from the reserve. Continue along Mansfield Road past the SSPCA centre on the left and at the next junction bear left to the Pentlands Regional Park car park at NT 166 639, also on your left. From the car park, walk south on the road to the reserve entrance, located about a hundred metres further on your right. A circular boardwalk leads onto the southern end of the reserve.

Place Names

Lenzie: possibly from the Gaelic *lèana*, meaning 'wet meadow'.

Blawhorn: 'warm hill', from the Gaelic *blàth*, meaning 'warm', and *càrn* meaning 'hill'. Local opinion suggests the name may derive from the coaching days when Blackridge was a midway station between Edinburgh and Glasgow and Blawhorn was used as a viewing point when a horn would be blown to signal the coaching Inn at Blackridge of approaching coaches.

Balerno: unusually for Midlothian, Balerno is from the Gaelic *Baile Àirneach*, meaning 'townland of the hawthorns'.

ENGLAND

Peatlands cover around eleven per cent of England, most of which is blanket bog in the upland chain running north from Derby to the Scottish Border. There are also blanket bogs among the hilltops in the Lake District and in the south-west on Dartmoor and Exmoor. Large areas of lowland raised bog have formed on flat low-lying land in the north-west of England, and on the Somerset Levels in the south-west. The East of England fenland once covered almost 4,000 square kilometres, but now most of the peat has wasted away through conversion to agricultural land, leaving a small handful of sites with fen habitat.

Few peatlands escaped people's need for fuel, with most showing signs of peat cutting stretching back thousands of years. More recent demands on peatlands include commercial mining to supply horticultural peat and the intensification of agriculture for livestock in the uplands and arable or pasture in the lowlands, often to the dismay of local people who valued the rich wildlife and plentiful resources of the undrained peatlands. The post-war expansion of the forestry industry saw many peatlands drained and planted with conifers.

Large parts of the uplands have been used for sport shooting as grouse moors over the last few centuries, with frequent burning as part of the management. In today's era of climate change awareness there is the perverse situation of tree planting, windfarms and peatland conservation all competing for the same land, resulting in less climate benefit than if all three were pursued without compromising one another.

What remains of England's peatlands is very much prized for its wildlife, with the lowland bogs and fens being some of the richest areas for plants and invertebrates in the country. Upland areas support breeding and wintering populations of our rarest and most threatened birdlife.

There can be no overstating the archaeological significance of England's peatlands to our understanding of past human activity and ecological change stretching back 10,000 years. The incredible preserving qualities of wet peatlands have revealed intimate details of people's lives as far back as the Bronze Age, allowing us to understand what they looked like, what they ate and wore, and how they interacted with the surrounding landscape.

Several of England's peatland sites are within Areas of Outstanding Natural Beauty and were among the first areas to be designated as national parks, in recognition of their importance for people to visit and enjoy the open spaces. More recently, the vital role of peatlands in tackling climate change, drinking water supply and flood management has led to nationwide efforts to restore and conserve the peatlands. Pioneering projects, supported by public and private funding, to repair past damage are providing inspiration and techniques that are being applied across the globe.

Fenland with bearded tits and marsh harrier

NORTHERN ENGLAND

Peatlands are physically and culturally at the heart of northern England, with blanket bogs adorning the Pennine Hills, the backbone of England that stretches from the Peak District through the South Pennines, the Yorkshire Dales and North Pennines up to the Cheviot Hills.

This northern upland chain has become a powerhouse for peatland conservation, with partnerships overseeing some of the largest peatland restoration efforts in Europe. These have become exemplars not only of the techniques for repairing peatlands in challenging remote conditions, but also of the support of local people (farmers, moorland managers and business owners) who have engaged with public bodies and environmental charities in a co-ordinated effort.

Surrounded by England's most populated, and formerly industrial, centres, the scenic uplands have long provided an escape from bustle, noise and pollution. In the 1930s the area became a flashpoint in the conflict between those seeking access to the hills and the landowners demanding privacy. Modern legislation and changed attitudes have embraced open access and the region is now a focus for rural recreation with well-marked routes and visitor facilities.

Britain's first long-distance trail, the Pennine Way, runs 429 kilometres from Edale to Kirk Yetholm. The route was conceived by the journalist and outdoor activist Tom Stephenson, who wrote in 1935 of his dream for a long green trail akin to the Appalachian Trail in the eastern United States. On 24th April 1965, the route was officially opened in the Yorkshire Dales village of Malham, with a nod to the protest known as the Kinder Mass Trespass, thirty-three years earlier to the day, when over four hundred people walked to the summit of Kinder Scout in protest at private landowners who had effectively closed the moors to the public. This symbolic event was part of a campaign that set in motion sweeping changes, with legislation eventually providing freedom of access and the formation of the national parks, the first being the Peak District National Park, designated in 1951.

In another reflection of the area's history, paving has been laid on parts of the Pennine Way since 1991 using flagstones from the floors of derelict mills in the West Pennines to overcome the erosion and treacherous conditions faced by those crossing the peatlands. In their heyday during the Industrial Revolution these mills were a major driver of the peatland degradation of the same moors that their flagstones later helped to protect.

The Border Mires in the west of Northumberland and east of Cumbria are a large group of deep peat lenses of blanket bog, raised bog and intermediate bog where the peat in deep basins has extended over the ridges to coalesce with adjacent peat domes. Much of the peatland lies within the Forestry Commission's Kielder Forest Park. Once part of a wider area of peatland and native wet woodlands, these remaining mires were isolated as large-scale drainage, and planting of conifers changed the surrounding land. Retaining some of the least damaged bog vegetation to be found in Britain, and still supporting rare bog species lost from the rest of England, the Border Mires have been a focus for conservation work since the 1960s. Action to block drains to restore the peatlands started in the 1970s. Since 1986, the Border Mires Committee has overseen conservation management work on the mires that has extended into high-altitude blanket bog, with three major phases of habitat restoration in the first two decades of the 21st century.

The exceptional quality of these mires and their significance in Britain is often overlooked, in part due to the sea of commercial forest now surrounding them. There are two Northumberland Wildlife Trust Reserves with good access that provide a good introduction to the Border Mires – Falstone Moss, south of Kielder Water (NY 708 860), and Bellcrag Flow (NY 775 725). A third trust reserve at Butterburn Flow (NY 660 758) is the largest of the Border Mires. It has no official access but can be viewed from the high point at the end of the surfaced road past Butterburn Farm.

The peatland plateau within the North York Moors National Park was one of England's largest blanket bog areas, but centuries of agricultural grazing and burning from sporting management, together with recent extensive fires, have left little of the natural bog habitat. Yorkshire Peat Partnership has been involved with landowners in a programme of work to attempt repair of the damaged peatland areas. Yorkshire Wildlife Trust's Fen Bog Nature Reserve (SE 857 982) in the east of the moors is a very rich valley mire offering views over the surrounding peatlands and a chance to see now rare peatland plants, such as bog sedge (*Carex limosa*) and white beak-sedge (*Rhynchospora alba*).

Packhorse bridge over the River Noe, leading to Jacob's ladder on the Pennine Way, Edale

PEAK DISTRICT MOORS

Map: OS 1:50,000 Sheet Nos 110, 119
Peatland Grid ref: SE 106 004
Access point: Moorland Visitor Centre, Edale (SK 123 856)

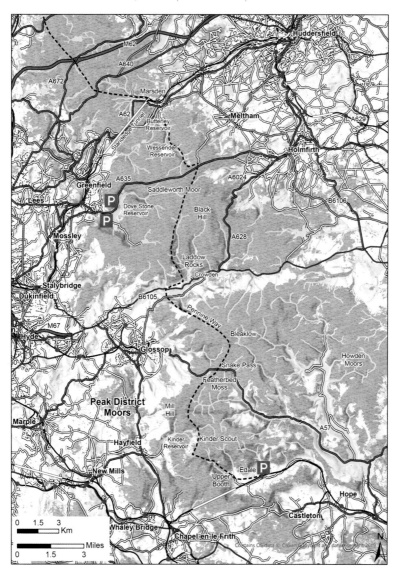

At the northern edge of the Peak District National Park is the Dark Peak, an area of high moorland plateaus and deep valleys with a mantle of blanket bog over grey gritstone.

Black Hill Trig point

Start of the Pennine Way at Edale

Contrasting with the limestone plateau and gentle farmland of the White Peak to the south, this is a region of extensive peatland that has been described by some travellers as a sea of treacherous bare, black peat, inhospitable, menacing and melancholic. Often shrouded by low cloud, a crucial factor in supporting the wet peat habitats, weathered stone tors appear as strange shapes among scree slopes, escarpments (edges) and wooded ravines (cloughs).

Alfred Wainwright, author and illustrator of several definitive guidebooks on fellwalking in the Lake District, complained that sinking up to his knees in peat while walking in the area was one of his most frightening experiences, but this is a description of the past when the peat had no protective mat of interwoven peat-forming vegetation. Today, the peatlands have undergone an incredible transformation. Conservation work is reversing centuries of damage from drainage, grazing and burning as well as pollution from the neighbouring industrial towns, that had left a desolate moonscape of blackened, eroding peat. An intense and co-ordinated effort to bring back the living bog vegetation has made a dramatic improvement. The results appear as if a carpet of greens, reds and gold had been laid over the scarred landscape. Moorland birds such as golden plover and dunlin are returning to breed, enriching the moors with their evocative calls.

Some of the best peatland experiences can be found along the route of the Pennine Way between the villages of Edale and Marsden, particularly around the summits of Kinder Scout, the highest point in the Peak District, Bleaklow, the second highest, and Black Hill. The blanket bogs that extend across the region have been divided into moors and individual mosses with intriguing names, such as Featherbed Moss, perhaps referring to cotton-grass. Bleakmires Moss is self-explanatory and Dove Stone Moss is aptly named as it is managed by the RSPB. The National Trust is a major landowner, with several thousand hectares of bog between Kinder Scout and Howden Moor making up an area known as the High Peak, all of which is now being managed with the aims of conservation and public access.

Saddleworth Moor and Winter Hill hit the headlines in 2018 when one of the largest wildfires to burn close to an urban area in the UK broke out, resulting in mass evacuations from local villages, businesses and nearby towns. Bleaklow Moor was also the site of a major fire covering over seven hectares in 2003, when plumes of smoke engulfed Manchester forcing the airport to close. Drained and damaged peatlands are vulnerable to reckless or accidental fires burning out of control, particularly as our spring and summer weather becomes warmer with climate change. Re-wetting the peatlands and restoring wet bog

Edale Rocks on route to Kinder Scout
Drawing: Dunlin

Flagstones from derelict mills leading across Featherbed Moss on the Pennine Way.

vegetation will greatly reduce the fire risk, as saturated peat doesn't burn. During the recovery period, while leggy heather and dry grasses remain, or in prolonged dry spells, fire safety is paramount and visitors are urged not to light fires and to report any they see.

The Peak District National Park is sandwiched between the huge conurbations of Manchester and Sheffield and attracts several million visitors a year. It takes careful management to prevent all that footfall from damaging the soft peatlands. Stone flags laid over parts of the Pennine Way are not always popular. They can damage the peat by compressing it, but they are effective in reducing the width of erosion from walkers and they guide people away from the most sensitive habitats. Despite the huge visitor numbers, the area is large enough to allow a peaceful escape from the cities.

Restoration of the peatlands has been co-ordinated through Moors for the Future Partnership, which began work in 2003. An impressive thirty-three square kilometres of eroded peat in the Peak District National Park and neighbouring South Pennines is on its way to recovery. Stabilising and revegetating the bare peat on such a huge scale in a remote area has required innovative methods and major engineering.

One of the biggest hurdles is the limited amount of sphagnum moss available to regenerate the peatland, most of it having been lost through heavy industrial pollution. Supplies have had to be brought in from other peatland sites, or locally sourced fragments have been specially propagated in nurseries and spread across the bare peat. Water companies in the region support the restoration, as drinking water for the surrounding cities flows from the blanket bogs. Damaged bogs release brown-coloured water requiring costly treatment. Repairing the peatlands reduces the colour in the water and in the long run saves money.

Travel Notes

The Moorland Visitor Centre, Edale at the start of the Pennine Way is a logical place to begin a peatland experience. There is a railway station in Edale served by frequent trains on the Hope Valley Line between Sheffield and Manchester. The Moorland Centre is less than half a kilometre from the station and has interactive displays explaining the work of Moors for the Future Partnership and the research team based there. There is also a small campsite.

A thirty-six-kilometre route between Edale and Marsden along the Pennine Way is a challenging and dramatic route taking in some of the best peatland areas. The Pennine Way starts in Edale village and leads out towards Upper Booth across pleasant grassy farmland that belies the challenging route ahead. The climb begins at Jacob's Ladder, once used as a packhorse route and named after an 18th-century farmer who set the stone path into the steep hill.

Follow the path to the top of Edale Rocks and, keeping a close eye on the cairns to show the route, continue to Kinder Downfall where on a windy day the waterfall can blow upwards. Cross the River Kinder a few metres from the falls and join the track heading north-west above Kinder Reservoir. Continue to the summit of Mill Hill, ignoring the Snake Path on the left to Hayfields and the track on the right to Ashop Clough. At the bottom of the hill a track of paving stones across blanket bog of Featherbed Moss leads to the Snake Pass and the A57, the main road between Sheffield and Manchester.

Cross the road and join the track to Bleaklow Head then downhill along the top of Clough Edge to the village of Crowden, a good overnight stop. From the village the Pennine Way climbs up a steep valley to the impressive Laddow Rocks, offering panoramic views and a daunting path alongside the near-vertical cliff edge. Paved stones again lead over blanket bog on the approach to the trig point at Black Hill summit, where sphagnum moss and cotton-grass have replaced bare peat thanks to conservation work.

To the north-east is the town of Holmfirth, famous as the location for the television series *Last of the Summer Wine*, but also for devastating floods when water rushes off the hills. The route then heads downhill to Dean Clough and then crosses the A635. Take a right turn along the road for a few metres and then left onto a small side road that continues as the Pennine Way. The route becomes picturesque as it leads towards Wessenden Reservoir. Turn right from here along the Kirklees Way passing Butterley Reservoir and onto Marsden village and the train station. Audio guides of other trails are available to download from the Moors for the Future Partnership website with accompanying maps and directions.

Place names

Peak: from an Anglo-Saxon tribe called the *Pecsaetan* ('peaklanders') or possibly from *peac*, an Old English word for 'hill'.

Kinder Scout: the origins are unclear, but Kinder might be related to Kintire or Kintyre (from the Gaelic *ceann*, meaning 'head' or 'point', and *tìr*, meaning 'country' or 'territory'). Scout may be a corruption of the Old English *sceot*, meaning 'shot' or 'shoot'. Scout hills are long ridges of rock, appearing to be 'shot out' horizontally. Alternatively, it may be from the Norse *scuti*, meaning 'overhanging rock'.

YORKSHIRE DALES AND SOUTH PENNINES

Map: OS 1:50,000 Sheet Nos 98, 102, 103, 104
Peatland Grid ref: Yorkshire Dales SD 890 820, South Pennines SD 992 283
Access point: Yorkshire Dales Park Centre (SD 874 898), National Trust Marsden (SE 046 117)

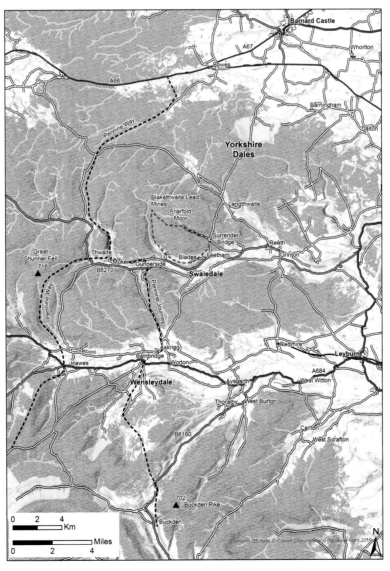

Grinton Moor looking across to the village of Reeth in the Yorkshire Dales

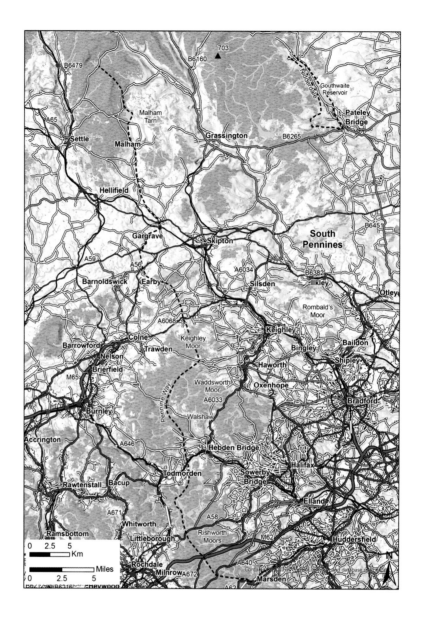

The Yorkshire Dales are characterised by their deep valleys and imposing hills. Drystone field boundaries and flower-rich hay meadows occupy the lower ground beneath steep heather-clad slopes with blanket bog on the tops. Much of the area is included within the Yorkshire Dales National Park, although the Nidderdale Area of Outstanding Natural Beauty to the east has been excluded despite having a similar wonderful landscape.

South of the national park, the South Pennines area has gentler, rolling hills and narrow valleys. The moors around Haworth provided the moorland setting and inspiration for Emily Brontë's book *Wuthering Heights*, the word 'wuthering' referring to strong winds. North-east of Keighley is Rombalds Moor, which contains Ilkley Moor, well known from the Yorkshire anthem 'On Ilkla Moor Baht 'at' – the local dialect for 'on Ilkley Moor without a hat'.

Greensett Moss from the flank of Whernside looking across Chapel le Dale to Ingleborough, Yorkshire Dales

The Dales, derived from the Norse word for 'valley', contain distinctive limestone rocks that form great pavements. These plateaus of bare and weathered rock are often found at the top of limestone cliffs, known locally as 'scars'. On the hill tops where blanket bog is found, the limestone is overlain by coarse millstone grits and shale. Some of the best blanket bog examples are found in Chapel-le-Dale and Ribblesdale in the west of the national park.

Most of the Dales' peatland is found in the north and east between Wensleydale (known for its cheese, first made in the 12th century by French monks who had settled in the region) and Swaledale (synonymous with James Herriot the veterinary surgeon and his popular series of books). The blanket bogs have been drained with thousands of ditches, or 'grips', dug into the peat in the years following the Second World War and facilitated by the invention of the Cuthbertson plough. This is the most intensely gripped peatland in England and the degradation has been furthered by sheep grazing and repeated burning on the widespread grouse moors.

In 2009 Yorkshire Peat Partnership was formed to co-ordinate restoration of the peatlands in the uplands of northern Yorkshire. A huge programme of blocking ditches and gullies, alongside government funding for land managers to reduce grazing and burning had

taken place on over 30,000 hectares by 2017, covering around a third of the region's peatland. The National Trust owns and is helping to restore large areas of blanket bog at Upper Wharfedale and Marsden Moor. The restoration work also includes grouse moors where private owners recognise the benefits of healthy peatlands for supporting red grouse by supplying water and food.

Water companies that manage the numerous drinking water reservoirs within the Dales have considerable interest in peatland restoration. Yorkshire Water has established a demonstration farm applying its 'Beyond Nature' philosophy at Humberstone Hub in Nidderdale. Moorland restoration and conservation are combined with a Swaledale sheep farming business, along with management of hay meadows and native woodland.

Another positive force for peatland conservation arises from the significance of peatlands for flood management. Most of the large rivers which have their source in the Dales flow eastwards through the Pennine fringe and into the Vale of York. In recent years there have been several major flood events in York as water rushing from the hills overwhelms the flood defences. River courses arising in the South Pennines have also seen flooding affecting smaller towns, such as Hebden Bridge and Pateley Bridge. Water coming off high ground after heavy rainfall travels much faster across damaged blanket bogs and eroded gullies than on restored and healthy peatland. The mosses and other peatland vegetation provide roughness that slows the water movement and reduces the downstream impact on flood defence systems.

Travel Notes

The South Pennine peatlands can be accessed by the Pennine Way. Marsden village is a good starting point, with a railway station and frequent trains from Huddersfield. The National Trust office beside the station has details on a number of short walks around the moors to the west of Marsden.

To reach the Pennine Way and onto Rishworth Moor, Wadsworth Moor and Keighley Moor, join the towpath alongside Huddersfield Narrow Canal and head west to the Standedge Tunnel and Visitor Centre. Cross the canal by the stone bridge and take the path through the Tunnel End Nature Reserve to join Waters Road.

Continue westwards along a small lane to Eastergate Cottage and take the waymarked path to the left through a gate and alongside the River Colne, nowadays a small stream. Cross the river at the packhorse bridge, marked as Close Gate Bridge. This was the route used to take wool from Yorkshire and cotton from Lancashire.

After crossing the bridge follow the track up the steep ridge, staying on the high ground and not using the tracks in the adjacent valley bottoms. This is the old Packhorse Road that leads past March Haigh Reservoir and on to meet the Pennine Way at the A640. For those prepared for a long hike, cross the road and take the Pennine Way north-west across Moss Moor and on towards Haworth and into the Yorkshire Dales.

Nidderdale can be reached by bus from the main rail station at Harrogate. The Nidderdale Way is a

Drawing: Golden plover

waymarked route from Pateley Bridge leading up to blanket bog on the moors around Scar House Reservoir. Lead and zinc mining dating as far back as the 12th century has left a distinctive mark on the area, particularly around Greenhow. Mining of these metal ores involved torrents of water being used to expose the mineral veins producing distinctive landscapes with mineral spoil heaps.

The Dales can be reached via the Pennine Way from Haworth with one of the first peatland attractions situated forty-eight kilometres north at Malham Tarn. Here you will find Tarn Moss, owned by the National Trust, a raised bog in a dramatic and popular setting. For a more leisurely journey to the Dales the picturesque Settle to Carlisle railway has a station at Horton in Ribblesdale to connect to the Pennine Way. The remote, bleak station further north at Garsdale has three bus connections a day to the main national park visitor centre at Hawes. There are many walks around Wensleydale from the town that was originally known as The Hawes, deriving from the Old Norse word *hals*, meaning 'neck' or 'pass between mountains'.

Coming into Wensleydale from the east, the heritage Wensleydale railway has regular services between Leeming Bar and Redmire. Trains are timed to link in with buses to Northallerton and the main rail network.

The Pennine Way reaches Swaledale at the village of Thwaite, with several walks to the surrounding moors. Some of the best moorland experience with evidence of Yorkshire Peat Partnership restoration work is around Gunnerside village, reached by bus from Thwaite or Hawes.

For a moderate four to five-hour walk start at Gunnerside village square, cross the bridge over Gunnerside Gill and take the marked path alongside the east bank of the river heading northwards up to Blakethwaite lead mine, first passing Bunton mine. At a four-way post (SD 939 012) continue ahead on the rising path signed Blakethwaite Dam. As the path drops down into the gill it passes Blakethwaite Smelt Mills and, a kilometre upstream, the Blakethwaite lead mines.

From here, take the track up the side of Friarfold Moor then join a path that heads south then east through more mine workings to reach Level House Bridge. Turn south-east alongside Hard Level Gill, pass the ruins of the Old Gang Smelting mills to reach Surrender Bridge. Turn right onto the road, cross the bridge and continue for four hundred metres. A signed path on the right leads across Feetham Pasture to the small hamlet of Blades.

Beyond the cottages is a tarmacked road. Turn right and follow it west as it narrows into a track and eventually a path leading uphill. The route continues across open pasture to reach a wall (SD 963 983). Fork left to leave the main path, which contours across the hillside, and descend steeply down a field that leads to a rough track dropping down into Gunnerside village.

Place names

Nidderdale: valley of the River Nidd, of Celtic origin and thought to mean 'brilliant'.

Swaledale: valley of the River Swale, Anglo-Saxon in origin, from *swalwe*, meaning 'rushing water'. The River Swale has a reputation for being one of the fastest rising flood rivers in England.

Pennines: possibly Celtic *pen* meaning 'hill', or the name may have been invented in the 1700s by Charles Bertram, possibly based on the name 'Apennines', from the mountains in Italy.

March Haigh: a march is an edge boundary and 'Haigh' is from Old Norse *haga*, meaning 'enclosure'.

NORTH PENNINES

Map: OS 1:50,000 Sheet Nos 86, 87, 91, 92
Peatland Grid ref: NY 804 379
Access point: Bowlees Visitor Centre (NY 906 281)

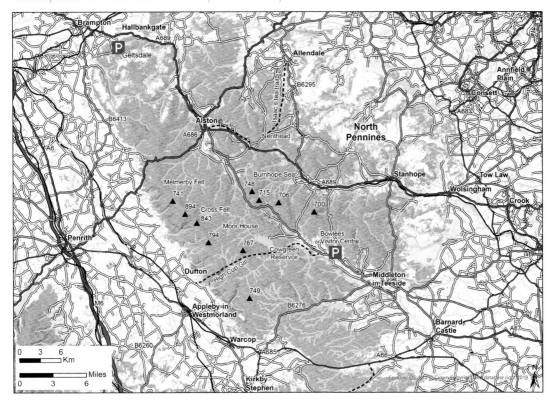

The North Pennines have a more remote and tranquil atmosphere compared with the South Pennines. This is an Area of Outstanding Natural Beauty (AONB), designated in 1988 particularly for its moorland scenery and containing the largest contiguous area of blanket bog in England. Blanket bog predominates on the top of the hills around Teesdale, Weardale and Allendale, with heath and hay meadows occupying the lower slopes. Distinctive terraced hillsides and flat hilltops are the results of alterations of thin limestone shale and thick sandstone being laid down and then eroding at different rates. The soft shales wear away more easily, producing contrasting bands of rock. So important is the geology that Britain's first European geopark was designated here and later was also recognised as a UNESCO Global Geopark.

Tynehead Fell, North Pennines

Peatland restoration work, North Pennines

Surprisingly less visited than the areas to the south, this is a region that has a wealth of natural and cultural heritage. For thousands of years rich seams of silver, and in more recent centuries lead, were mined around Alston and Nenthead, England's highest village. Evidence of the huge scale of mining activity remains today, with numerous earthworks, abandoned mines and mills as well as the homesteads of the miners who later turned to farming. Among the many archaeological finds from the peat here have been 4,000-year-old horns of aurochs (Bos primigenius), an extinct species of very large wild cattle.

Much of the land is now used for sheep farming and grouse shooting. As well as supporting red grouse, the North Pennines hold over eighty per cent of England's black grouse (Lyrurus tetrix), a bird that lives on the moorland fringe alongside woodland. The RSPB has a nature reserve at Geltsdale in the north-west corner of the North Pennines. The land includes two working hill farms with livestock, and conservation management is helping maintain and enhance the lower-lying meadows and woodlands as well as the heath and blanket bog on the higher ground. Hen harriers have successfully bred here in recent years after an absence of over a decade.

The North Pennines also holds one of England's largest and earliest National Nature Reserves, Moor House-Upper Teesdale, containing blanket bog, rare arctic-alpine plants such as spring gentian (*Gentiana verna*) and an excellent example of juniper (*Juniperus communis*) woodland.

As with many other blanket bog areas in England, there has been considerable damage to the peatlands from past drainage, grazing and burning, leaving large exposed bare areas of eroding peat. It is worth noting that although the Pennine Way runs alongside the edge of Moor House NNR and all managed burning on the reserve ceased with its purchase in the early 1950s, there has never been a wildfire on the reserve, except when a managed burn on a neighbouring estate ran out of control and burnt a marginal section of the reserve. Since 2006, the North Pennines AONB Partnership has run a Peatland Programme to co-ordinate restoration on over 35,000 hectares of peatland, including blocking over 10 million metres of drains and revegetating large areas of bare peat with sphagnum.

Travel Notes

The Pennine Way connects the market town of Middleton-in-Teesdale with the town of Alston, giving excellent views of the peatlands west of the River Tees. The nearest railway station is Darlington, with bus connections to Middleton-in-Teesdale via Barnard Castle. From the west, buses connect Alston with Penrith railway station.

A short distance from the village of Dufton on the Pennine Way is a small station at Appleby on the Settle to Carlisle railway, with trains to Leeds or Manchester. Above Dufton is the sweeping U-shaped valley of High Cup Gill, a remarkable geological feature set among large expanses of blanket bog and heathland. The Pennine Way leads eastward from Dufton through the blanket bog and the Moor House-Upper Teesdale National Nature Reserve to a visitor centre in a former chapel building at Bowlees near the famous High Force waterfall on the River Tees.

Isaac's Tea Trail is a long-distance walking route linking Alston with Allendale. The trail begins from Isaac's Well in Allendale. The trail is named after a 19th-century tea seller and fundraiser called Isaac Holden. Originally a lead miner, failing health and the decline in the mining industry led Isaac to embark on selling tea door to door among the farming and mining communities throughout the remote hamlets.

The RSPB Reserve at Geltsdale (NY 588 584) is five kilometres from the Brampton Junction railway station on Carlisle to Newcastle line. Head east along the A689 to Hallbankgate (where Robert Stephenson's famous Rocket steam locomotive ended its working life) and onto the RSPB Stagsike Cottages information centre.

Place names

Geltsdale: valley of the River Gelt, from Old Norse *gelt* meaning 'gold', in reference to the colour.

Alston: previously *Aldenby*, from Old Norse, a person's name 'Halfdan'.

MIDLANDS

This region is low-lying and flat land across the central part of England. The underlying geology is, predominantly, a red clay laid down as mud and silt 250 million years ago, during the Triassic era. Post-glacial, waterlogged hollows and shallow lakes became inundated with fen vegetation and many of these later formed sphagnum-dominated lowland raised bogs.

In the west, the Marches Mosses, covering the North Shropshire, Maelor and Cheshire plains, is a unique landscape with the largest concentration of natural lowland lakes and wetlands in England and the highest density of ponds in Europe. At the end of the last Ice Age the retreating ice deposited glacial moraine, consisting of sands, gravels and clays, in places up to fifty metres thick, giving the area its characteristic undulating terrain. This wide mix of wetland habitats includes open water, swamp, fen, alder carr, marshy grassland and peat bog. The Whixall group of mosses together form one of the largest lowland raised bogs in Britain.

To the east, the Humberhead Levels are a large expanse of flat, low-lying land at the western end of the Humber estuary and occupying the southern half of the area of the glacial Lake Humber that formed during the last Ice Age and later filled with clay sediment. As sea levels rose the numerous rivers across the flat plain backed up, creating the waterlogged floodplain fens and bogs, comparable to the fens of East Anglia.

The relatively accessible deep peats were extensively exploited for centuries to be converted to agricultural land and cut to provide fuel, animal bedding and more recently horticultural products.

Both east and west areas are comprised of a number of large remnant sites. Long appreciated as open, uncluttered landscapes, they still retain a wealth of peatland wildlife. Extensive restoration works have been required to re-wet, and in some cases, revegetate the damaged peatland. One of the most inspiring aspects of these projects is the ambition to extend management to the surrounding landscapes. The aim is to recover some sense of the huge former scale of mixed wetland habitats, incorporating sympathetic farming activity, and to help return viable populations of some of our rarest wildlife.

Richard Lindsay measuring peat depth at Hatfield Moors

HUMBERHEAD PEATLANDS

Map: OS 1:50,000 Sheet Nos 111, 112
Peatland Grid ref: Thorne SE 730 160, Hatfield SE 705 060
Access point: Moorends (SE 704 161), Crowle (SE 758 140), Hatfield (SE 693 068 and SE 683 049)

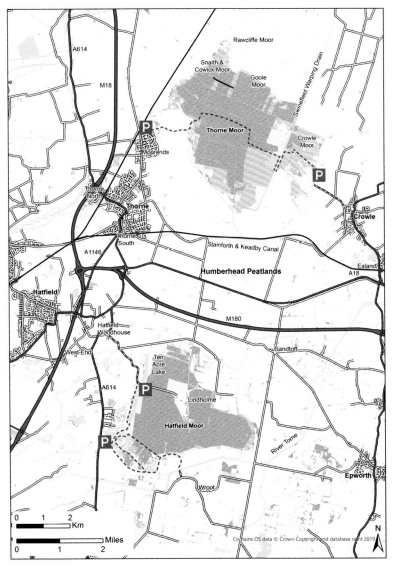

Thorne and Hatfield Moors, situated to the north-east of Doncaster, are remnants of an extensive former wetland, now mostly converted to agricultural land, that occupied the

Peatland restoration area on former peat mining works at Hatfield Moors

flat, low-lying Humberhead Levels. Thorne Moor (1,900 hectares) and Hatfield Moor (1,400 hectares) together form the largest remaining complex of lowland mire in Britain. Having been perceived as wasteland fit only for drainage, dumping and development, these moors have been the subject of intense conservation battles and are now central to one of England's most visionary wildlife recovery schemes.

Thorne Moor lies north-east of the town of Thorne and west of Crowle and comprises several 'moors' defined by parish boundaries: Thorne Waste, Rawcliffe Moor, Snaith and Cowick Moor, Goole Moor and Crowle Moor. Once part of the same peatland complex, Hatfield Moor is now separated by a corridor of arable farmland containing the Stainforth and Keadby Canal and the M180. Hatfield Moor lies over glacial sand and moraines, the largest of which forms Lindholme Island in the centre of the bog.

Thorne and Hatfield Moors are owned and managed by Natural England as the Humberhead Peatlands National Nature Reserve. The Lincolnshire Wildlife Trust owns a 188-hectare nature reserve on the edge of Thorne Moor at Crowle Moor. The ecology of the area is expertly captured in a series of papers published by the Thorne and Hatfield Moors Conservation Forum, a voluntary group instrumental in the survey, conservation and management of the moors. Thorne and Hatfield Moors boast the richest variety of insect species of any peatland in the UK and are home to species found nowhere else in the country, such as the mire pill beetle (*Curimopsis nigrita*), nicknamed the 'bog hog',

Drainage ditch on Thorne Moors

whose tiny female lines her burrow in the peat with sheets of moss, and the Thorne pin-palp beetle (*Bembidion humerale*), known only from the fossil record and thought to be extinct before being discovered alive on the moors in 1975.

Birdlife here is also abundant with crane, bittern and marsh harrier among the other peatland specialists. Occupying the overlap between the breeding ranges of northern species and southern species, the area used to provide the unusual opportunity to see birds such as whinchat (*Saxicola rubetra*) and nightingale in the same locality; sadly, since 2000 the nightingale has declined and no longer breeds here. There are almost a hundred breeding pairs of nightjars (*Caprimulgus europaeus*), a nocturnal bird known locally as the 'gabble ratchet'. At dusk and in the evening the males sit camouflaged on a branch or on the ground making a distinctive churring sound followed by an eerily silent flight display, interrupted only by a croaking call and whip-like wing claps.

Once part of the ancient royal hunting ground of Hatfield Chase, the peatlands were renowned for the diversity of game and wildlife they supported. During the 17th century in the era of agricultural improvement, priorities changed, and King Charles I asked the Dutch engineer Cornelius Vermuyden to drain the wet, marshy land. Local people unhappy at the disruption and loss of common land mounted legal challenges, rioted, and took direct action to reflood the land over the course of a century before the unrest settled.

The beginning of the 19th century saw commercial exploitation of the peat, mainly as animal litter, as a material to absorb urine and faeces, particularly in horse stables, with the demand lasting until the internal combustion engine replaced horsepower. Peat was cut by hand and by the 1890s removed by horse-drawn barges along a network of canals that required water levels to be kept relatively high, giving some respite to uncut parts of the bog. Despite the damage, mid-19th century observers still described Thorne Moor an active 'quaking' bog with a peat dome six metres in depth, that could rise over two metres in a wet winter before falling again in the summer. The British Moss Litter Company added a narrow-gauge railway, initially using horse drawn wagons on the rails to take the peat to the processing mills on the edges of the moors. It was not until the 1960s that horses were replaced by railway engines. The chequerboard appearance of cuttings and baulks from peat removal in this era can still be seen.

Another of the products from the peatland was 'bog wood', used for fuel and fence posts. The wood was the preserved remains of a buried forest of oak, Scots pine and birch that grew before climate change and fire, possibly deliberately caused by Bronze Age people, led to the trees' demise and inundation by peat. A few Scots pine reputed to be of natural origin remain today on Hatfield Moor among planted exotic conifers. A wooden Bronze Age trackway discovered on Thorne Moor in 1972, and a Neolithic one on Hatfield Moor in 2004, along with the preserved ecology of the ancient woodlands, are part of the rich archaeology of the peatlands.

In 1854 an act of parliament allowed an agricultural improvement called 'warping'. River and tidal waters were channelled through 'warping drains' to flood the cut-over and subsided peat and deposit layers of silty soil 'warp' as much as a metre thick. Remains of the Swinefleet warping drain can be seen on the east of the moors. Much of the land around Thorne and Hatfield Moors was converted by warping to arable and grassland, a practice still used today on peatlands in Japan but no longer in Britain. At the close of the century the remaining peatland had noticeably subsided by over three and a half metres, and it was reported that whilst formerly only the top of Goole church spire was visible across the moors from Crowle, by the late 1890s the whole steeple was visible. The warping work stopped in 1916 but peat cutting continued into the 20th century, albeit at a lesser scale, as coal replaced the use of peat for fuel and demand for horse bedding declined with the advent of the car.

In the mid-20th century, the moors became the focus for new developments including plans to dump ash from the Drax coal-fired power station. In 1963 the moors were obtained by the horticulture division of Fisons, which mechanised the cutting of peat blocks. Amateur naturalist Peter Skidmore, who was employed as Keeper of Natural History at Doncaster Museum, was one of the leading figures in recording the wildlife importance of the moors and championed their conservation. At the same time, a group known as 'Bunting's Beavers' took direction and began dam building to block drains in 1972. With attention widely drawn to the plight of the peatlands, the next two decades saw legal conservation designations and management agreements gradually being applied, but not without some backward steps.

Hatfield Moor, which had remained predominantly uncut, although partially drained in the 19th century, was subjected to mechanised peat cutting in the 1960s. Then in 1987 Fisons introduced 'milling', a highly destructive form of peat extraction. After a period of public campaigning, the last peat milling ended on Hatfield Moor in 2006. Later threats arose in 2012 with the development of Tween Bridge windfarm on the northern edge of Thorne Moor and later additional turbines were constructed to the south.

The start of the 21st century heralded a new phase for Thorne and Hatfield Moors, now known as the Humberhead Peatlands. Largely secured in conservation hands and with European Union and UK Government funding, restoration management is taking place on a grand scale. There are considerable challenges in repairing what one visiting scientist described as a sort of Frankenstein's monster, because its complex patchwork of drains and cuttings makes water management extremely difficult. One solution was the introduction of a pumping station on Thorne Moor at the Swinefleet drain using an Archimedes' screwpump to lift water to where it was needed.

Two decades on and the recovery is remarkable. Cotton-grass and moss grow on once featureless bare peat expanses and wet, bog vegetation is returning to drained areas that had been inundated by scrub and heath. The nightjar population which had taken advantage of these drier habitats is being catered for with new heaths and woodland habitat on the edge of the peatland where it belongs. The most exciting part of the recovery is the formation of the Humberhead Levels Partnership with its vision for connecting the wetlands of Thorne and Hatfield Moors with those of the Humber estuary, Derwent Ings and the Aire Valley. The aim is to create an amazing landscape not just rich in biodiversity but supporting a new generation of farming and visitor experiences, while reducing flood problems and addressing climate change.

Travel Notes

The nearest main rail station is Doncaster, with regular train connections to the town of Thorne which has two railway stations, North and South, and to Crowle on the east of Thorne Moor.

The Peatlands Way, with its nightjar logo, is an eighty-kilometre circular route linking the communities of the Humberhead Peatlands and traversing the peatlands of the Thorne, Crowle and Hatfield Moors. The route is linked to the Trans Pennine Trail at the New Junction canal.

The Peatlands Way can be accessed from the Delves Fishponds (SE 681 137) in Thorne, adjacent to the Thorne North Station and where there is a café and toilets. From here the Way leads to the Moorends Recreation Ground (SE 699 158) at the end of Grange Road. To the side of the gate a public path leads round the disused Thorne Colliery, crossing the main road to the pit. Take the signposted track over the metal footbridge and onto Thorne Moor, where there are information boards and a marked route leading east to Crowle Moor. Moorends can also be reached by bus from various stops in Thorne.

Alternatively, if accessing from the Crowle side, an unclassified road leads north-west from the village to a parking area and waymarked routes onto Crowle Moor and with access to Thorne Moor along the Peatlands Way.

Hatfield Moor can be reached by bus from Doncaster to Hatfield Woodhouse for access from the north-east of the site, via Remple Lane to Ten Acre Lake on the reserve or to Wroot village for access from the south. From the main street in Wroot, turn north along the Peatland Way to the River Thorne to cross a footbridge then turn left onto Hatfield Moor. Turn right after four hundred metres to the edge of the old commercial peat extraction area. Turn left and follow the track to the south-eastern edge of the peat workings to a flooded lake and the start of the Moor Bank track (SE 695 038) where there are hides and a boardwalk.

Follow the Peatlands Way through the reserve for the next six kilometres to a tarmac lane and the Ten Acre Lake with a lakeside viewing platform. To the right is a private road to Lindholme Hall, with the Peatlands Way bearing left then heading north, becoming Remple Lane, which leads to the village of Hatfield Woodhouse.

Place names

Swinefleet: Old Norse *Sweyne's* channel. One interesting suggestion is that it refers to a small channel where pigs regularly uprooted old roman coins known locally as 'swine pennies'.

Crowle: from a Dutch settlers' name for 'cattle pens', or from *crowle*, which means 'to creep' in Lincolnshire dialect. Local legend describes the area as a safe area of high ground to crawl up and avoid being drowned in the bog.

Thorne: Anglo Saxon, meaning 'the place where the thorns grow'.

Snaith: from an Old Norse word *sneith*, meaning 'a detached piece of land'.

Cowick: Old English 'cow enclosure'.

Hatfield: Old Norse *heithr*, meaning 'heath' – a tract of open uncultivated land.

Goole: Old English *goule*, meaning 'channel', or possibly 'outlet drain'.

MARCHES MOSSES

Map: OS 1:50,000 Sheet No 126
Peatland Grid ref: SJ 490 365
Access point: Morris's Bridge (SJ 493 355), Wem Moss (SJ 471 341)

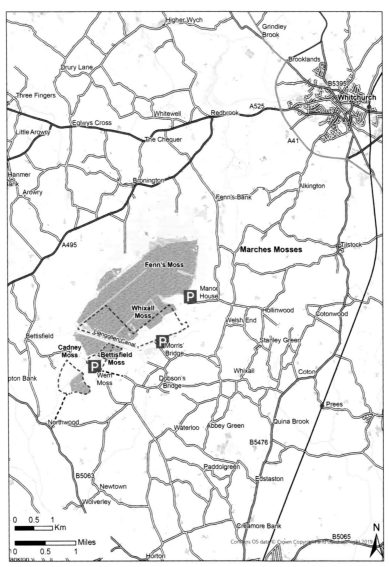

The Marches Mosses are a group of lowland raised bogs in the plains of the English-Welsh borderland, the Marches, near Whitchurch in Shropshire. The largest group, known

Aerial view of Fenn's Moss

collectively as the Whixall Moss complex, is sub-divided into parish units with Fenn's, Bettisfield and Cadney Mosses in Wales, and Whixall Moss and Wem Moss in England. This was once a single large peatland unit that has been fragmented and reduced by conversion to agricultural land over the centuries. The peatlands are set within a region known as the Midland Meres and Mosses, one of Europe's great wetland landscapes comprising shallow lakes (meres) and raised bogs (mosses) extending out from Cheshire into Shropshire, Staffordshire and parts of North Wales.

Despite bearing the scars of thousands of years of human intervention with peat cutting, agriculture and more recently forestry, the mosses have retained an incredible wealth of peatland wildlife. Over twenty-nine species of damselflies and dragonflies, including the rare white-faced darter, have been found here. The seven-centimetres wide raft spider, which can be seen standing on open water among the sphagnum carpets, is one of the more obvious of the rare bog spider residents. During the summer months, the mosses resonate with the calls of curlew and skylark. There is also a good chance of spotting a hobby (*Falco subbuteo*), one of our smallest birds of prey, whose main food is dragonflies. Along the scrubby edges of the bogs, nightjars can be seen at dusk hunting for moths. In winter there is still plenty to see, with hen harrier and short-eared owl silently quartering the peaty expanse.

Fenn's, Whixall and Bettisfield Mosses form a large peatland National Nature Reserve mostly owned by Natural England and Natural Resources Wales, with nearby Wem Moss also a National Nature Reserve, owned by the Shropshire Wildlife Trust. The reserve has a fascinating history of human intervention, much of it evident today. Three preserved Bronze Age and Romano-British bog bodies discovered in the late 19th century on Whixall Moss may indicate the significance of the area as a tribal territorial boundary, at a time when peatlands were used for ritual human sacrifice.

For many centuries, local people cut peat from around the edges of the peatland for fuel, leaving the centre largely untouched. During the era of agricultural enclosures in the late 18th and early 19th centuries, much of the bog surface was drained and the outer areas converted to agricultural land. Further drains were installed in 1804 as part of the construction works for the Llangollen Canal that separates the south-west corner of Whixall Moss and Bettisfield Moss. Promoted by iron and coal industrialists to connect the region to the River Mersey and the River Severn, this section of the canal was originally called the Ellesmere Canal and its construction across three kilometres of deep peat posed a challenge for the engineers William Jessop and Thomas Telford.

Even with drains lowering the peatland water table, the canal had to be constructed on a floating raft. The route required constant maintenance to build up the bank walls to avoid flooding using a resident team of workers known as the Whixall Moss gang, who continued operating till the 1960s. The solution to preventing the canal sinking further was to underpin the whole section with steel piling. Another engineering impact involving

Fenn's Moss showing restoration using bunded cells to retain water levels

drainage was the construction in 1863 of the Oswestry, Ellesmere and Whitchurch railway, crossing three miles of Fenn's Moss. Disused since the 1960s, the railway line was floated on the bog using heather bales, birch logs, sand and flagstones.

Commercial peat cutting began in the mid-19th century on Fenn's and Whixall Moss and nearby Wem Moss. Initially providing livestock bedding and litter, the peat cutting was taken over by the War Office in 1915 to provide bedding for cavalry horses. Sphagnum mosses were also harvested as wound dressings for soldiers. After the war, the peatland was worked by commercial companies providing peat for horticultural use. By 1989 the scale and intensity of mechanised peat cutting led to environmental campaigns against the damage and this resulted in the centre area of the Fenn's and Whixall Mosses being bought by the statutory agencies and an end to peat cutting there by 1990.

Bettisfield Moss, which had avoided much of the peat extraction and supports the oldest deepest peats, became inundated by pine seedlings from nearby plantations to such an extent that they provided a harvest of Christmas trees up until the 1960s, when they were abandoned and grew into a woodland. Cadney Moss in Wales was largely converted to agricultural land and forestry.

Conservation work over the last few decades has seen intensive work to clear scrub and invading trees from the peatlands and a comprehensive programme of blocking drainage ditches to help re-wet the bog surface. Some of the former peat workers were employed in this work, using their skills in navigating machinery over the bog surface

Bettisfield Moss following restoration work to remove trees and scrub

and their knowledge of the drain network. There have been recent signs of success with recovery of sphagnum mosses and the first breeding snipe returning to Whixall Moss in 2019 after an absence of over twenty years. The significance of this news is all the greater as there are only ten breeding pairs of this important wetland bird in the whole of Shropshire.

Conservation works on the peatlands also involved deep piling being installed along the canal in 1990 to prevent contamination of the bog. There are plans to establish wetland habitat, including fen, willow and alder carr woodland around the peatlands, to help buffer them from the drainage effects of the adjacent agriculture land. In 2001 the woodlands on Bettisfield Moss were cleared along with the removal of invading tree saplings across the reserve, accompanied by drains being blocked. This work initiated restoration, but because the surface of the peat was so damaged by peripheral drains and tree growth, it was not sufficient to create the permanently wet surface conditions needed for the sphagnum moss carpet to regenerate. Cell-bunding – a technique developed on Natural England's National Nature Reserves in Cumbria in which impermeable low peat 'bunds' are used to create a network of water-holding cells – has recently been completed over Bettisfield Moss and other parts of the site. These cells ensure that rainwater is retained on-site long enough for the 'acrotelm' to re-form, and the site to once again start growing peat.

Travel Notes

There is a rail station at Prees, six kilometres from the reserve, with a two-hourly service here northbound from Shrewsbury and southbound from Crewe. The nearest bus stop is three kilometres from the reserve at Coton. From either Prees or Coton the reserve can be reached by joining the Shropshire Way marked trail that leads through part of the reserve. The reserve is also near Route 45 of the Sustrans National Cycle Network that leads south-west to the reserve from the rail station at Whitchurch.

For Fenn's and Whixall Mosses, three interlinking circular trails pass through the reserve and take in the Llangollen Canal. Routes start from the Morris's Bridge car park (SJ 493 355) on the Llangollen canal near Whixall. A history trail starts from the Manor House National Nature Reserve office (the former peat factory) car park in Whixall (SJ 505 366).

Bettisfield Moss can be accessed along a figure-of-eight track from the World's End car park (SJ 480 348) situated at the end of the second road on the right, west of Dobson's Bridge, Whixall. The trail can also be reached from the east along the Shropshire Way Route 23.

Nearby Wem Moss is a Shropshire Wildlife Trust Reserve (SJ 471 341). From Northwood, take the north-bound track from the eastern edge of the village. The track leads over a footbridge, through some trees and onto the reserve.

Place names

Wem: from the Saxon *wamm*, meaning 'a marsh'.

Whixall: Old Norse, from a person's name, *Hwittuc's*, and *halh*, meaning 'firm ground in a wet area'.

Cadney: Old Norse person's name 'Cada's island'.

NORTH-WEST ENGLAND

Peatlands in the north-west corner of England range from the tranquillity of Cumbria's estuarine plains along the Solway Firth and the blanket mires of the Lake District and Lancashire's Forest of Bowland to the major conurbations of Liverpool and Manchester that retain only fragmented remnants of once vast raised bogs.

Cumbria's peatlands include the South Solway Mosses, some of the best examples of raised bog in Britain, as well as raised bogs to the south of the Lake District National Park, which developed in valleys draining south into Morecambe Bay. Good examples for visiting include the Cumbria Wildlife Trust Foulshaw Moss reserve near Witherslack, with a boardwalk and viewing platform. Forest planting and associated drainage in the 1950s and 1960s damaged almost the whole site but repairs are being made by the trust to restore the wetland. Nearby Roudsea Mosses at the head of the Leven estuary and Duddon Mosses National Nature Reserve hidden away at the top of the Duddon estuary offer a lowland mix of raised bog and fen. The lovely village of Grizebeck provides a panoramic view of the Duddon mosses and the Lakeland Fells.

The Lake District mountains and hilltops are adorned with blanket bogs, many of which have been damaged by agricultural drainage and forestry in the past. Large areas are now being restored including Bampton Common, which sits above the RSPB Haweswater Reservoir in the eastern Lake District, and Shap Fells, where work is supported by the water company United Utilities. Across Cumbria, the Cumbrian Peat Partnership has put more than 4,900 hectares of peatlands into restoration in the last ten years. Within the Lake District valleys there are numerous fens known as valley mires with a rich plant community benefiting from the varied geology supplying mixes of acidic and base-rich water. They support many uncommon species including flat-leaved bog-moss (*Sphagnum platyphyllum*), the most alkali tolerant of the sphagnum mosses.

The Bowland Fells form a distinctive upland plateau east of Lancaster in the Forest of Bowland. The area shares features with the neighbouring Yorkshire Dales, with blanket bog-clad summits covering millstone grits. The bog habitat is heavily degraded in parts with bare peat, open gullies and peat haggs. It remains an important site for the region's iconic bird of prey, the hen harrier. The peatlands also hold the headwaters for several important rivers that provide drinking water for Lancaster and nearby towns. Restoration work over the last ten years and into the future as part of the government-funded North of England peat project is aimed at improving the habitat and water quality.

Lancashire's lowland raised bogs have been reduced to three per cent of their former area, with the remaining 7,600 hectares spread across the region as small fragments vulnerable to pressures from surrounding agriculture land. Chat Moss between Manchester and Liverpool was once England's fifth largest raised bog but has been greatly reduced over the centuries. In the 19th century, Stephenson's railway connected the two main cities and ran right across the moss, nearly bankrupting Stephenson until he hit upon the idea of floating the railway on fascines of brushwood – on which it still runs today. At the time of Stephenson, the railway enabled peat to be cut and transported to Manchester, with the return wagons carrying the city's night waste to be dumped as fertiliser for agricultural improvement of the bog.

Lancashire Wildlife Trust is working with partners to create a pioneering 'carbon farm' at Winmarleigh, part of the Trust's Winmarleigh and Cockerham Moss Nature Reserve. This former commercial peat extraction site is being re-wetted and as part of the work sphagnum moss is being grown to aid habitat recovery and also to pilot the production of moss to supply the horticulture market as a sustainable replacement for peat, which is just ancient sphagnum.

Bowness Common

SOUTH SOLWAY MOSSES

Map: OS 1:50,000 Sheet No 85
Peatland Grid ref: Bowness Common NY 191 595, Wedholme Flow NY 220 528
Access point: Solway Wetlands Centre (NY 197 615), Wedholme Flow car park (NY 238 539)

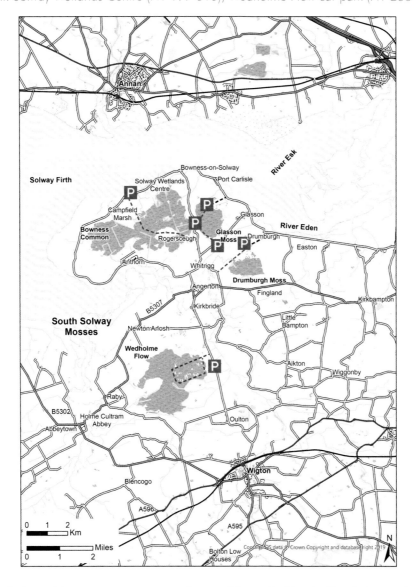

The South Solway Mosses are a group of raised bogs in North Cumbria on the southern edge of the Solway Firth and centred around Kirkbride to the west of Carlisle. Set within the

Wetland pool, Bowness Common

Glasson Moss observation tower

large, flat expanse of the Solway coastal plain, these bogs form a complex with estuarine habitats of salt marsh and wet pasture that offers a rich array of wildlife. The Solway Coast is designated as an Area of Outstanding Natural Beauty with vast, unbroken vistas across the border to Scotland in the north and to the Lake District in the south. Remains of the western end of Hadrian's Wall mark this region as the frontier of the former Roman Empire.

The peatlands here support some of the largest and best examples of raised bog habitat in England. Bowness Common, Glasson Moss and Drumburgh Moss are located along the coast with the second largest site, Wedholme Flow, situated just south of the village of Kirkbride. All are designated as National Nature Reserves and most are owned by

Natural England. There has been a long history of peat cutting in the area, and many of the bogs show scars from early peat cutting since medieval times through to the more extensive mechanised peat removal of modern times.

Bowness Common, previously referred to as Bowness Moss or Flow, has survived largely unscathed by peat extraction apart from evidence of hand cutting for fuel around the peatland edge and drainage works in the south of the site. What remains is the largest area of raised bog habitat in England, although partially damaged by past drainage, burning and grazing. It was originally part of an even more extensive peatland that included the nearby Glasson Moss, now separated by agricultural land and a road.

The relatively flat terrain here is interrupted by the Rogersceugh drumlin, a twenty-five-metre-high mound of glacial deposit that provides spectacular views of the surroundings. Previously supporting an agricultural small holding, the area is now being restored to peatland habitat including fen and willow carr by the RSPB, who manage it as part of their Campfield Marsh Reserve. The Solway Wetlands Centre on a separate part of the reserve, where peatland meets coastal farmland and saltmarsh, near Bowness on Solway, is a focal point for visitors.

To the east of the Rogersceugh drumlin is the route of a disused railway constructed in the late 1860s connecting Bowness to Annan in Scotland by a viaduct across the Solway. The line between Whitrigg and Bowness cut across the peat which was up to fifteen metres deep. Extensive drainage works across the peat saw water running in river-like streams from either side of the line for many weeks and resulted in the peat surface dropping by about two metres. This peat was still not stable enough for trains, and bundles of faggots had to be laid to form a floating track. After the Solway viaduct disaster in 1881, when large ice blocks carried on the tides crashed into the supporting structure and required extensive repairs, the line struggled economically and was closed in the 1920s. There were newspaper reports of Scots rashly walking across the disused viaduct to obtain drink in England before it was dismantled in the 1930s. British Rail sold the land in 1990 to Natural England, which has since been blocking the old drains and reinstating fen habitat as a precursor to eventual bog restoration.

Glasson Moss, a raised bog to the east of Bowness Common, has been impacted in the past by drainage, fires and peat cutting, particularly in the area of Whitrigg Common in the southern section. The central area retains a relatively intact mire surface, and thanks to recent restoration works the peatland is showing great recovery. Management includes improving areas of carr scrub woodland around the bog edge to support one of our threatened bird species, the willow tit (*Poecile montanus*).

Drawing: Hen harriers

Looking north over Bowness Common to Bowness on Solway

Nearby Drumburgh (pronounced drum bruff) Moss is owned by the Cumbria Wildlife Trust and having also suffered from drainage and peat cutting is being restored through ditch blocking and removal of invasive trees and shrubs. The Trust has also reintroduced white-faced darter, a rare dragonfly that breeds in deep bog pools with floating sphagnum moss used for egg laying by adults and as a refuge by larvae.

To the south of Kirkbride, Wedholme Flow has experienced an intense history of peat extraction. As far back as the 16th century, there was large-scale peat cutting to provide fuel for the Cistercian monks who manufactured salt by evaporating seawater at nearby Holme Cultram Abbey. However, it was industrial-scale 20th-century peat works that caused the most extensive damage. The early cutting was done by hand using Dutch techniques by men wearing wooden footboards to stop them sinking into the peat. From the 1960s the bog was more intensively drained and milled, with large areas of peat exposed and extracted for use as gardening compost.

Since 2007 the site has been in the hands of Natural England, which has been managing water levels necessary for peatland restoration and has reintroduced sphagnum moss species from donor sites on Bowness Common. Despite the extensive damage to the middle section of the site from peat workings, large sections of relatively intact raised bog habitat remain in the north and south of the site, with important populations of peatland species including colonies of large heath butterflies.

Travel notes

The nearest railway stations are Wigton, 9.5 kilometres south of Wedholme Flow, and Carlisle, twenty-one kilometres east of Bowness Common. Regular bus services connect with the main villages beside the main access points to the peatlands.

Bowness Common can be reached from the Solway Wetlands Centre at the RSPB Campfield Marsh Reserve on the main coast road, 2.5 kilometres west of the village of Bowness-on-Solway. The centre is a restored 19th-century barn, part of the original North Plains Farm, which houses exhibits and offers refreshments and toilet facilities. A marked track and boardwalk lead south to viewpoints across the moss at Rogersceugh Farm.

Glasson Moss is north of the B5307, near the villages of Glasson and Whitrigg. The northern access point is a small car park accessed via a track from the cottage and Glasson caravan park two kilometres north of Glasson village. An impressive wooden viewing tower marks the start of a boardwalk and gravel trail that leads to the western access point on the Kirkbride to Bowness-on-Solway road. Whitrigg Common at the south end of Glasson Moss can be reached by a track from a parking layby on the Whitrigg to Glasson road at NY 239 593.

Glasson and Bowness-on-Solway are both on National Cycle Route 72.

The Cumbria Wildlife Trust Reserve at Drumburgh Moss is reached by an 800-metre track from Drumburgh village to a small car park on the north edge of the bog. Marked trails lead out to a wooden viewing platform.

Access to Wedholme Flow is from a car park on the eastern side of the bog just off the Wigton to Kirkbride Road at NY 238 539. Marked trails lead to a viewpoint over the peatland restoration works.

The Solway Coast AONB authority has established a number of cycling routes, including a thirty-one-kilometre ride around Wedholme Flow from Kirkbride. The route can also be accessed from Moorhouse, two kilometres north of Wigton train station.

Place names

Wedholme Flow: Old English *wathe*, meaning 'hunting' and *holmr*, a small island in the middle of the *flow* or 'marsh'.

Glasson: Anglo-Scandinavian *glaise*, meaning 'a small stream'.

Rogersceugh: Roger's *sceugh*, from Old Norse *haugr*, meaning 'a low hill or mound'.

Bowness: from Old Norse *boga*, meaning 'a bow' or 'curved' and *naess*, meaning 'promontory'.

EASTERN ENGLAND

The Fens coastal plain in eastern England covers 40,000 square kilometres in the counties of Cambridgeshire, Lincolnshire, Norfolk and Suffolk. About 4,000 years after the end of the last Ice Age, Britain was separated from the rest of Europe by rising sea levels that also slowed the flow of fresh water from the uplands to the Wash estuary, causing rivers to flood and wetlands to form. Fens, bogs and lakes replaced the earlier oak woods that have been preserved in the peat for thousands of years, giving rise to the bog oak frequently exposed where the peat is eroded or cut.

For several thousand years people lived around the edges of the Fens and on the islands of higher ground. Livelihoods were based on pastoral farming, fishing and fowling, and the harvesting of reeds and sedge for thatch and baskets. Some of the best examples of Mesolithic settlements have been found preserved in peat at Must Farm quarry and Flag Fen, including Bronze Age boats spanning a period of about a thousand years. These are now being preserved at the Flag Fen archaeological park on the eastern outskirts of Peterborough along with a reconstructed Bronze Age village.

Attempts at drainage for agricultural improvement had taken place since Roman times but were largely unsuccessful. In the Middle Ages numerous powerful landowning monasteries began cutting peat as a commercial enterprise, selling the peat for fuel to nearby towns and cities. In Norfolk, 14th century records for Norwich Cathedral show the priory kitchen alone used 400,000 peat turves a year, much of it extracted from peatlands that later flooded to become the lakes and fens of the Norfolk Broads.

The era of agricultural drainage that started in the 17th century and continued more effectively in the 18th and 19th centuries resulted in the loss of over ninety per cent of the once vast Fen peatlands. What little of the original habitat and its wildlife remains is scattered among a few pockets of land, mostly protected as nature reserves. Surprisingly, these hold an incredible wealth of species, with Cambridgeshire and Lincolnshire holding over twenty-five per cent of the UK's most threatened and rare plant and animal species.

Exploring new ways of farming with crops that benefit from wet conditions is part of a grand new vision for the fenlands, alongside extensive restoration of wetlands with fens, bogs and wet woodland aimed at giving new and lasting benefits to the region. The realisation of this vision is starting with a network of fen rehabilitation projects around Cambridgeshire's Holme Fen and Woodwalton Fen as well as Wicken Fen and Chippenham Fen, together with the RSPB's Lakenheath Fen work at the River Little Ouse in Suffolk.

Holme Fen Post showing peat shrinkage, with the top cap marking ground level in 1848

CAMBRIDGESHIRE FENS

Map: OS 1:50,000 Sheet Nos 142, 154
Peatland Grid ref: Holme Fen TL 205 895, Woodwalton Fen TL 230 840, Wicken Fen TL 555 700
Access point: Holme Fen Posts (TL 203 894), Woodwalton Fen (TL 234 849), Wicken Fen (TL 563 705)

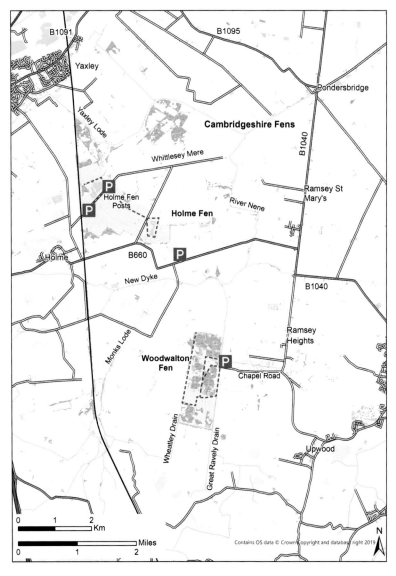

A vast flat plain stretches out across Cambridgeshire, east of the A1(M) between Peterborough and Huntingdon, to the Wash estuary on the North Sea coast. The patchwork

Drain in Woodwalton Fen

Stands of common reed (*Phragmites australis*), Woodwalton Fen

of arable fields bounded by a network of straight-lined ditches and rivers represents one of the most profound ecological changes imposed by the agricultural revolution in this country. The clue to this dramatic transformation is in the black peaty soils beneath the modern crops of England's breadbasket.

What existed four hundred years ago was an immense peatland landscape of bogs, fens, reedbeds, carr woodland and open water that provided the region with its name, the fenlands. Drainage schemes from the 17th century onwards have resulted in over ninety-nine per cent of the peatland habitats being destroyed. The surviving remnants of that once great fen exist as a handful of sites of which Wicken Fen, Woodwalton Fen and Holme Fen are the largest. From these vestigial refugia the fate of some of our rarest and most threatened wildlife hangs in the balance while work starts on an inspirational series of projects to reinstate a fenland landscape on a grand scale.

Holme Fen and Woodwalton Fen, both National Nature Reserves, lie to the south of Peterborough at the western edge of the fen basin. Although situated only a few kilometres apart they have very different characteristics. Both originated near the south-west shores of a great open water body covering over 1,000 hectares and forming England's largest lowland lake, Whittlesey Mere. Once a haven for waterfowl and a venue for regattas and ice-skating competitions, the lake succumbed to the 19th-century agriculture improvements in one of the last major drainage activities in the fenlands. The largest steam pump of the time was deployed to remove the water and within a few years the lake and all its rich wildlife was replaced by wheat fields.

Holme Fen was a raised bog before it was drained, and unlike other parts of the fens the sphagnum peat was not removed, making this one of the few sites to have its original peat surface. Perhaps too wet to remain suitable for arable farming, the area became overgrown, eventually forming birch woodland with additions of planted oak in the mid-20th century to provide gamebird cover and material for charcoal production. Now in the hands of Natural England, the combination of wetland and woodland has a rich array of wildlife with over five hundred species of fungi alone.

Other notable features within the woods are the Holme Fen Posts, two cast iron monoliths standing among the trees. Their incredible story reveals one of the most dramatic consequences of the 19th-century drainage of the fens, that of subsidence. In 1848, when plans were being made to drain Whittlesey Mere, local landowners were concerned about the surrounding peat drying out, shrinking and lowering, with the risk of flooding and possible inundation from the sea. A long post was driven down through seven metres of peat to the clay substrate, with its top cap at ground level to allow any changes in peat depth to be monitored. In just over 170 years the ground has dropped to leave an incredible four metres of post now visible. Holme Fen is now reputed to be the lowest point of land in Britain at around 2.7 metres below sea level. Here, as in the rest of the Fenlands, the still subsiding peat soils require a massive and constant effort through river engineering, ditch management and pumping to prevent farmland and settlements from flooding.

Conserving Holme Fen is a challenging task as the wetland is perched above surrounding farmland that is shrinking faster due to the combined effects of drainage and wastage as the cultivated peat erodes to be blown away in the wind and washed down the ditches. Attempts at keeping the nature reserve wet through blocking drains have helped, but more costly engineering solutions would be required with no guarantee of success, particularly with competing arable farming interests and a changing climate.

A more holistic and lasting solution has come through the Great Fen project, covering over 3,600 hectares of surrounding land, including nearby Woodwalton Fen. The project aims to halt peat wastage through raising water tables, creating wetlands and introducing new forms of farming with crops that don't require drainage. The woodland habitat at Holme Fen will undoubtedly recede as the raised bog recovers. As one of England's largest birch woods, this relatively recent artefact of man-made drainage of the peatlands is a sad indictment of the limited state of our woodland heritage elsewhere. Putting in place new native woodlands in those naturally drier parts of the surroundings will return a landscape of wetlands and woodlands not seen for two centuries. The combination of benefits arising from the Great Fen project, for wildlife, tourism, farming and flood prevention, makes this a visionary and inspirational plan.

Woodwalton Fen, managed by Natural England, is one of the oldest nature reserves in Britain. At the end of the 19th century, Charles Rothschild, a city banker and keen entomologist alarmed at the loss of wild areas, began to acquire land for conservation, including the purchase in 1910 of Woodwalton Fen. A bungalow that Rothschild built as a base for his wildlife studies still stands there today. Once a raised bog associated with Whittlesey Mere, the sphagnum peat was removed up until the early 20th century, to be used as fuel in the nearby Ramsey Heights brick-making kilns. Reedbeds, willow and alder took over the remaining peat soils giving the wonderful mix of fen habitats we see today, with bitterns and marsh harriers regularly occurring here as well as occasional common crane, a species that disappeared from the fens four hundred years ago.

Wicken Fen lies north-east of Cambridge and south of Ely, one of the few areas of elevated land in the Fenlands with its landmark cathedral. The fen was purchased by Rothschild in 1899 and shortly after donated to the National Trust, making it their oldest nature reserve. The site survived the 19th-century agricultural conversion and was used as a source of sedges for thatching roofs, a practice which has been recorded here from as far back as the early 15th century. Many of the 9,000 species of plants and animals found

Aerial view of Woodwalton Fen showing surrounding drained agricultural land with lower water table

here benefit from this ancient cycle of sedge cutting and regrowth, so it is continued today as part of the conservation management.

The site also contains the last surviving wooden wind pump, dating from 1912, now used to help pump water into the reserve to maintain high water levels. Since 1946 the Trust has been extending the site through reinstating fen and wet grazing meadows on neighbouring farmland. As a result, hen harriers and marsh harriers have returned, with bitterns breeding in 2009 for the first time in over eighty years. The success of this work has led to the development of a hundred-year Wicken Fen vision to restore wetlands, extending 5,300 hectares south and east towards Cambridge.

Travel Notes

Peterborough is the nearest railway station for Holme Fen, with buses to Ramsey St Mary's stopping at Holme village. The reserve is immediately north of the B660. It is best accessed from Holme village following the sign from the Holme to Yaxley Road which leads over a railway crossing to a number of parking places at the entrance to the Holme Fen Posts (TL 203 894), with marked trails through the woodland and onto the reedbeds.

Woodwalton Fen is midway between the railway stations at Peterborough and Huntingdon with buses to the nearby village of Ramsey Heights. Parking is at the western end of Chapel Road/Heights Drove Road, beside the Great Raveley Drain, with non-vehicular access over the new bridge into the reserve (TL 235 848).

There is an information board, car park and the start of a network of trails into the Great Fen on the north side of the B660 Ramsey St Mary's to Holme Road (TL 222 876).

Wicken Fen is south of Wicken village, nine miles from the train station at Ely and can be reached by National Cycle Route 11. From the A1123 in Wicken, take the Lode Lane to the Wicken Fen Visitor Centre.

Place names

Ely: originally Isle of the Eels, from the Anglo-Saxon *eilg*. Eels were a common feature of the fens before agricultural drainage and were even used as a form of currency.

Wicken: the place name element *wic* is from Anglo-Saxon, meaning 'a farm' or 'group of huts'.

Holme: Old English from the Old Norse *holmr*, meaning 'island' or 'meadow field in marshland'.

SOUTH ENGLAND

Extensive peatlands in southern England are largely concentrated in the south-west, where Atlantic influences bring wetter weather. Blanket bog has formed on the higher ground of Dartmoor, with smaller, more damaged areas in Exmoor and Cornwall. A large area of peatland called Goss Moor in mid-Cornwall occupies a valley basin at the headwaters of the River Fal supports a variety of bog and fen habitats much of which was heavily disturbed by tin mining from the 11th to 19th centuries but remains important for wildlife.

The Somerset Levels, a low-lying coastal plain, once held large areas of fen and lowland raised bog, most of which were subjected to intensive drainage in the 18th and 19th centuries, with intricate networks of channels and pumps created to prevent flooding of the subsided peatland now managed as agricultural land. In the 20th century, a large area west of Glastonbury was commercially mined for horticulture peat then the land was sold to Natural England in 1994. The Avalon Marshes partnership has brought together the government bodies and environmental charities, including the RSPB and the Somerset Wildlife Trust, which together manage thirteen square kilometres for conservation, primarily as open water and reedbeds. The visitor centre at Avalon Marshes (ST 425 414) between Shapwick and Westhay provides a focal point for visiting the various wetland nature reserves in the area.

The Greensand ridges of the Blackdown Hills in Devon and Somerset support numerous small 'springline mires' found particularly in the upper slopes of the valleys. A good example of this habitat can be found at the Somerset Wildlife Trust Yarty Moor reserve at the headwaters of the River Yarty.

Further east, the only significant remaining large area of peatland is in the New Forest National Park, home to over ninety distinct valley mires including some of the most intact examples in Europe. The largest of these is Cranes Moor, once a raised bog but cut over for fuel and now supporting mostly fen habitat.

The uplands of Cornwall and Devon were formed by huge granite intrusions and are poorly draining. This gave rise to the blanket bogs and valley mires of Bodmin Moor and Dartmoor, by contrast with the free-draining sedimentary rocks of Exmoor, which has thinner peat soils as a result.

Exmoor and Dartmoor have been designated as national parks, with excellent visitor facilities and guidance on the history and wildlife value of the peatlands. The huge importance of these peatlands for downstream water supply to the surrounding populations encouraged South West Water, together with local partnerships, to support restoration work across all three moors starting in 2011. Following the success of these initial projects, further work is being taken forward with funding through Natural England, the Environment Agency and government agriculture payments to farmers.

Upper West Dart River, with Crow Tor and ruins of tin miners' huts

DARTMOOR MIRES

Map: OS 1:50,000 Sheet Nos 191, 202
Peatland Grid ref: SX 614 752
Access point: Postbridge Visitor Centre (SX 646 788), Combestone Tor car park (SX 670 718)
Princetown Visitor Centre (SX 590 734)

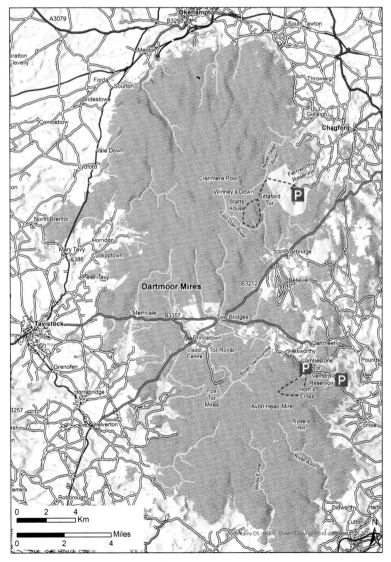

Dartmoor in South Devon is a vast upland area supporting Britain's largest expanse of granite. The hard, impervious rock in a region of high rainfall creates ideal waterlogged

Molinia covered bog, Upper West Dart River, Dartmoor

conditions for peatlands to form and gives rise to the distinctive boulder outcrops, known as tors, that are perched on the hilltops and weathered into natural carved monuments.

Rich in both cultural and natural heritage, this dramatic landscape was designated as the Dartmoor National Park in 1951. Beneath the veil of this tranquil remoteness is a complex history of widespread human interaction stretching back thousands of years. Today, further interventions are being made to remedy some of the unfortunate consequences of Dartmoor's industrial past.

The Dartmoor hills are cloaked in blanket bogs. There are also numerous river and stream channel bottoms containing a type of fen officially termed valley mires and locally referred to as 'quakers' or 'featherbeds'. The peatlands extend over 315 square kilometres between Okehampton in the north and Ivybridge in the south, with Princetown in the centre. The peat soils have preserved a wealth of archaeological information, demonstrating the area's importance for humans over the last 10,000 years, with abundant prehistoric settlements, burial tombs, monuments and ritual sites. The main period of peat formation in Dartmoor began around 4,500 years ago in the late Neolithic and into the early Bronze Age, when forests were cleared for cultivation and later abandoned as the climate became wetter.

Shallow blanket peat around Combestone Tor, West Dart River

Dartmoor is rich in tin ores and the history of tin extraction and processing is inextricably linked to that of the peatlands. Before the era of tin mining, early medieval tin was sourced from rivers and streams, and the dams constructed by tin workers, or 'tinners', led to the expansion of many of the valley mire fens in the waterlogged soils. The smelting of tin required a heat source and peat from Dartmoor, dried and turned in to charcoal, was regarded as a valuable fuel for the tin industry throughout Devon and Cornwall. Later, the Victorians developed large commercial hand-cut peat extraction sites, most of which soon failed economically in the face of cheaper coal. The huge scale of peat cutting over at least six hundred years can be seen as scars all across the Dartmoor landscape. Place names with the elements 'black' and 'turf' associated with workers' settlements and access routes can also be found throughout the region.

The widespread peat cutting together with 19th and 20th century intensive sheep farming, drainage and burning has left over ninety-nine per cent of Dartmoor's peatlands in a modified state, much of it severely degraded and with eroding bare peat in places. Natural bog vegetation can be found in patches across the region and some of the former hand-cut sites are showing signs of sphagnum recovery.

A feature of the blanket bogs in Dartmoor is the abundance of purple moor grass (*Molinia caerulea*), a species that grows naturally in bogs as single stems peppered throughout the sphagnum mosses. In damaged peatlands, especially those that are frequently burned, the moor grass takes over and in winter the dead leaves form a bleached white carpet across the landscape.

The deterioration in peatland habitat in the last century has been accompanied by declines in typical peatland breeding bird species such as golden plover and the world's most southerly population of dunlin. Peatland restoration work in recent years has seen an increase in dunlin numbers. Dartmoor still holds onto some of its rare and threatened species, including the black darter dragonfly, southern damselfly (*Coenagrion mercuriale*) and bog orchid in the valley mires and several sphagnum mosses on the blanket bogs. Occasional tall hummocks of golden orange Austin's bog-moss, now very rare in England, were a feature here but have not been recorded since 1967.

Most of the open unenclosed ground on Dartmoor is 'common' land which is privately owned, but other local 'commoners' have certain rights including grazing their livestock. These rights have been regulated since medieval times, but the origins of common land may go back even further. Current Dartmoor Commons legislation allows for public access to the commons.

In 2010 a five-year project explored the potential to restore Dartmoor's damaged peatlands at Winney's Down, South Tavy Heads and Flat Tor Pan through blocking old drains to raise the water table and restore peatland vegetation. Much of the work was funded by South West Water, responsible for the region's drinking water and concerned about the impact of the degraded peatlands. Nine of Devon's main rivers and forty per cent of the population's drinking water is sourced from Dartmoor peatland. The success of the restoration works has led to further government funding for peatland restoration as part of a South West Peatland Partnership, including neighbouring Cornwall and Exmoor.

A large section of Dartmoor is used by the Ministry of Defence as a training ground and access is strictly controlled during periods of live firing. The area has been used by the military for over two hundred years and there are ammunition craters and vehicle tracks scattered over the area. Today the management is less intrusive, but still damaging to the peatland. The military are undertaking conservation surveys to assist in the restoration of these areas.

Some of the best remaining of the Dartmoor peatlands are in the headwaters of the River Avon (or Aune) and the River Swincombe to the south-east of Princetown, where the national park has its flagship visitor centre. A few kilometres from the centre, Tor Royal Bog is the only raised bog in Devon or Cornwall. Nearby Fox Tor Mire is a valley mire set among a large expanse of blanket bog and reputed to be the inspiration for the renamed Grimpen Mire in Sir Arthur Conan Doyle's novel *The Hound of the Baskervilles*. Dr Watson is introduced to the mire by Jack Stapleton, a local entomologist, who describes it as 'a bad place' and where a false step 'means death to man or beast'. Stapleton also explains that islands of high ground within the mire are 'where the rare plants and the butterflies are'. Dr Watson later gives an account of the 'foul quagmire' with its 'rank reeds and lush, slimy water-plants' sending up an 'odour of decay and a heavy miasmatic vapour'.

Other folklore around the Dartmoor Mires includes the legend of the witch Vixana, who lived in a cave by Vixen Tor and could call up mists to lure people to their death in the bog at the foot of the tor. Travellers even up till the end of the 19th century believed in Dartmoor's pixies and feared becoming disorientated on the moors and being lured or 'pixie led' into the bogs. Carrying a piece of bread or wearing a coat inside out were considered protection from the mischievous creatures.

West of the Fernworthy Reservoir the blanket bogs around the East Dart River provide an opportunity to see the impact of former peat cuttings and the restoration works at Winney's Down, which have left few traces of the repairs other than the positive sign of much wetter ground.

Travel Notes

For the south-east of Dartmoor the nearest rail station is at Newton Abbot, with buses to Holne village for Aune Head Mire and Fernworthy Reservoir for Winney's Down.

The National Park Visitor Centre at Postbridge can also be reached by bus from Newton Abbot and has information on a number of trails across the moorlands.

For Aune Head Mire, a detailed route map is available from the Devon Wildlife Trust (Combestone Tor to Aune Head Mire). The route includes open grassland that can be very wet. From the village of Holne, head two kilometres north-east to Venford Reservoir and a further two kilometres to the Combestone Tor car park (SX 670 718). A grassy path across the road leads south-west past some stone enclosures and up to the top of Holne Ridge. Just as the path descends turn right to head along Sandy Way.

Follow the Sandy Way west for 1.5 kilometres to a broad, shallow valley and the head of the River Avon. In good weather there are spectacular views of the surrounding peatlands. The return route retraces Sandy Way past the point where it was first joined and then, once the old mineral workings are visible, leave Sandy Way, curving left, and follow the gentle descent past Horn's Cross and onto the Combestone Tor car park.

Winney's Down is approached from Fernworthy Reservoir and a detailed route map is available from the Devon Wildlife Trust. There are buses to the reservoir from Newton Abbot. From the reservoir car park (with parking fees) there is a path alongside the reservoir through woodland. Follow the sign to a bird hide and continue through a gate to the road and turn right to cross the Sandeman Bridge.

Follow the tarmac track to its end and the main forest gate. A stone track leads through the plantation going straight over any crossroads to the edge of the forest. A path leads south-west across open ground towards Sittaford Tor and passing the Grey Wethers stone circles on the left.

From the Tor there are excellent views across the peatlands. From here there is open boggy ground requiring considerable care. The route continues south-west to the right of the derelict stone wall, with the finishing point, at the ruin of Statts House, visible in the distance. There is a water crossing which can be difficult in wet weather.

Statts House was built as a shelter for the 19th-century peat cutters who worked in this desolate remote area. Near the hut are the remains of a track marked on the OS maps as a 'peat pass'. This is one of many tracks cut through the peat in the late 19th century by a local huntsman, Frank Phillpotts, who organised their construction across deeper parts of the mire to aid fellow huntsmen and other travellers. Marker posts and stone slabs with brass plates still present today indicate the start of the passes. Their upkeep proved too difficult as the peat eroded and collapsed around them and they were eventually abandoned.

The return journey can retrace the route to Sittaford Tor or head south to the East Dart River, past old tin workings and a nearby valley mire. Follow the river south-east to a waterfall and take the route of a dried-out water channel, or 'leat', following the contour away from the river. Follow the leat and after crossing the Winney's Down Brook head to a drystone wall. Turn left following the wall back to Sittaford Tor and from there you will reach the reservoir car park.

In the west of Dartmoor peatlands, the Princetown Visitor Centre can be reached by bus from Plymouth rail station. The centre occupies the old Duchy Hotel, where Sir Arthur Conan Doyle stayed while writing part of *The Hound of the Baskervilles* and visiting nearby Fox Tor Mire.

Place names

Dartmoor: 'moor of the oak grown stream' from the Brittonic root word *dar*, meaning 'oak' or possibly from *dwr*, meaning 'water'.

Fox Tor Mire: A tor is a prominent outcrop of rock derived from Old English *twr*, meaning 'a tower'.

Sittaford Tor: Anglo-Saxon *sith*, meaning 'a path'.

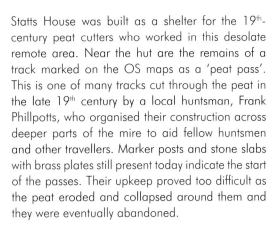

WALES

Wales is famed for its valleys and high mountains, from Snowdonia in the north to the Cambrian Mountains of Mid Wales and the Brecon Beacons in the south. There is a saying in North Wales that if the country were flattened out it would be bigger than England. Influenced by the wet Atlantic climate, these ranges support the largest expanse of blanket bog in southern Britain, with deep peat covering four per cent of Wales.

The lowlands of Wales, including the island of Anglesey, Pembrokeshire and Carmarthenshire, are important for their fens and raised bogs. There is a distinct cultural identity in Wales reinforced by the retention of the native Welsh language. The rugged landscape, difficult to access, resisted change and influence from outside until the Industrial Revolution, with its demand for the rich coal reserves of South Wales. Peatlands feature strongly in the Welsh language as these are areas where people would dig peat for fuel. *Mawn* ('peat') is extended to *mawnog* for 'peatbog', most often as places where peat was cut, and *cors* is associated with the wet vegetation of bogs or fens. Digging peat in Welsh is *lladd mawn* which literally translates into 'killing peat'. The Welsh name for dunlin, one of the typical breeding wader species of peatlands, is *pibydd y mawn* ('peat piper'), reflecting the bird's reedy trilling call.

Much of the country has a long history of livestock farming, with breeds such as the Welsh black cattle dating back at least 2,000 years, and the various small, hardy Welsh mountain sheep that are at home among the bogs and high pastures. My early career with the RSPB in Wales in the mid-1980s took me to the Denbigh Moors, *Mynydd Hiraethog*, to study the dramatic declines in breeding waders, which were of concern to environmental bodies. What we witnessed was a consequence of agricultural change and forest expansion that took place across the UK. The promised benefits of this expansion into the peatlands were little realised as the land remained too wet in many places for sheep or trees to be economically worthwhile or to justify ongoing subsidy. By the end of the 20th century, payments for high sheep numbers were stopped and forestry planting on peatland was dwindling rapidly.

Despite the decline in livestock and forestry on bogs their legacy on the peatlands remains, with devastating loss of bog vegetation often associated with inundation by purple moor grass, erosion, gullies and peat haggs over much of the blanket bogs and raised bogs, accompanied by a huge reduction in breeding waders. Curlew, golden plover and dunlin breeding numbers have dropped by as much as ninety per cent in the last forty years, with similar collapse of other peatland bird populations due largely to the loss of invertebrate food supply as the peatlands deteriorated.

Far from being bleak, the future for peatlands in Wales looks promising as government funding focuses a programme on restoring and conserving the many wonderful remaining upland and lowland sites, still teeming with wildlife. The earliest recovery projects in the afforested peatlands and open moorland are showing that nature can respond positively, with wildlife returning to even the most degraded areas. The motivations for this work go beyond wildlife to address flooding and climate change as well as heralding a new era in which farmers are rewarded as managers of the peatlands in their healthy state.

The three Welsh national parks, Snowdonia, the Brecon Beacons and the Pembrokeshire Coast, provide an excellent opportunity to explore the mountains and lowlands of Wales and learn about the peatland projects taking place. Upland sites managed by the RSPB at Vyrnwy and by the National Trust at Ysbyty Ifan offer a chance to experience the peatlands, with long-distance routes such as Glyndŵr's Way, which crosses the northern Cambrian Mountains, and the Monks' Trod along the Elan Valley. Numerous smaller nature reserves, many managed by the Wildlife Trusts in Wales, provide places to see fens and raised bogs across the country.

Greenland white-fronted geese, Dyfi estuary, Powys

NORTH WALES

Dominated by Snowdon, the highest mountain in England and Wales, the Snowdonia National Park is the largest in Wales and includes over a third of the country's peatland, mostly in the Migneint and Berwyn Mountains to the south of the park. Over half the population of this region speak Welsh, and there are signs and public information in Welsh and English. The Welsh name for the Snowdonia area is Eryri, previously thought to be derived from the Welsh for 'eagle', *eryr*, but now understood to be from the Latin *oriri*, meaning 'Highlands'.

The dramatic mountains are fringed by a mosaic of rivers, streams and wet pastures marked out by drystone walls, with oak and birch woodlands in an intimate mix. Farms here traditionally managed areas of *ffridd*, derived from 'woods in a field'. Livestock was moved in spring from lower ground to higher enclosed fields of scrub and woodland before going to the open grazing of the mountain tops. Many of the birds of the uplands utilise both the open blanket bogs and these lower farmed habitats during their breeding season, making the combination of features very important for conservation.

The National Trust's Ysbyty Ifan estate near the picturesque village of Betws y Coed and spanning the Machno, Eidda and Upper Conwy Valleys, is a large upland area of farmland and moorland leading to blanket bog on the high tops. There are several walks offering a chance to see the typical relationship of farming, forestry and peatlands in this mountainous area of the Migneint. The Trust has been working with other partners to restore areas of drained and eroding blanket bog, with further work proposed to help with downstream flood management. Further south-east in the Berwyn mountain range, just outside the national park boundary, is the RSPB nature reserve at Lake Vyrnwy, where the surrounding hills have been the focus for a blanket bog restoration project.

Contrasting with the uplands of North Wales is the island of Anglesey, whose fascinating geology has led to the island being declared a UNESCO Global Geopark. The low-lying terrain supports fens that are important for alkali-loving plants which benefit from the calcium-rich water draining from the surrounding limestone. No fewer than three of the larger valley-head fens have been designated as National Nature Reserves and have limited or good access facilities.

Blanket bog, Hafod, Lake Vyrnwy

ANGLESEY FENS

Map: OS 1:50,000 Sheet No 114
Peatland Grid ref: Cors Erddreiniog SH 470 820, Cors Bodeilio SH 500 775, Cors Goch SH 497 813
Access point: Cors Erddreiniog (SH 458 821), Cors Bodeilio (SH 506 772), Cors Goch (SH 504 816)

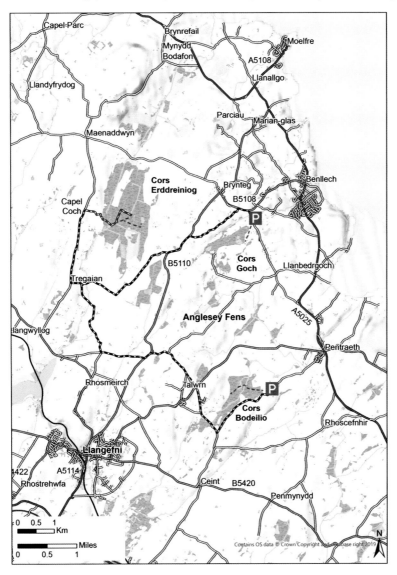

The low-lying island of Anglesey (Ynys Môn) is situated off the north-west coast of Wales and is separated from the mainland by the Menai Strait. The anglicised name is thought

Fly orchid, Cors Bodeilio, Anglesey

Cors Goch, Anglesey

to have an Old Norse origin, meaning 'Ongull's island', with the Welsh name being derived from the Brittonic *enisis mona* and later referred to by the Romans as *Mona*, but with no agreement as to its meaning.

The Anglesey Fens (Corsydd Môn) are the second-largest expanse of fen habitat in England and Wales, after the Norfolk Broads fens. They are particularly important as some of Europe's best examples of calcareous and alkaline fens, in headwater contexts, fed by underground waters that drain from the surrounding limestone rocks. The main concentration lies east and north-east of Llangefni in a series of shallow valleys, with three main sites designated as National Nature Reserves, the largest being Cors Erddreiniog along with Cors Bodeilio and Cors Goch.

The fens are home to reed and sedge-beds and important areas of low and open short-sedge fen dominated by black bog rush (*Schoenus nigricans*), and a range of slender sedges, including lesser tussock sedge (*Carex diandra*) and slender sedge (*Carex lasiocarpa*), with extensive areas of fen meadow dominated by blunt-flowered rush (*Juncus subnodulosus*) and purple moor grass. The great fen sedge is the stand-out feature of the areas of sedge fen and was once used for capping thatched roof ridges. The black-bog rush tussocks form an intricate micro-landscape of tussocks and wet runnels, and these provide the primary locus for such treasures as the fly orchid (*Ophrys insectifera*) and the narrow-leaved marsh orchid (*Dactylorhiza traunsteineri*) which depend on the alkali-rich waters.

The combination of large rush and sedge pastures, open water, acid heath, scrub and hazel woodland provide homes for an incredible diversity of insects, including some of our rarest dragonflies, damselflies and butterflies. In spring and summer, the air is filled with birdsong from grasshopper warblers, sedge warblers, reed buntings (*Emberiza schoeniclus*) and the lesser whitethroat (*Sylvia curruca*).

A long history of human influence on the fens has seen them managed largely sympathetically in the past, with traditional livestock and pony grazing along with the harvesting of reeds and sedges for animal bedding and thatch. The switch to more modern agriculture had a double negative impact on the fens; the land around them was drained, cutting off the supply of the calcium-rich water, as well as allowing the input of nutrients from animal slurry and fertilisers. The areas were seen as wasteland and Cors Goch was even proposed as a dumping ground.

In more recent times, livestock production has fallen and the fens have been abandoned, with scrub and woodland invading the drying out peatlands. In 2009 a five-year programme of fen restoration, funded by the EU LIFE programme, the largest of its kind

Fen meadow, Cors Bodeilio

Cors Bodeilio

in Europe, began the task of re-introducing traditional livestock grazing, removing and re-profiling damaged enriched soil, clearing scrub and adjusting water levels to support the fen vegetation. The work secured the support of local farmers to maintain the fen by reintroducing ponies and cattle grazing at levels that would keep down the competing grasses and scrub while allowing the fen plants to grow.

One of the additional benefits of this wildlife conservation work has been an improvement in the quality of the water that leaves the fens and enters the Llyn Cefni Reservoir. Further restoration and sympathetic management are being encouraged across a wider area to help reduce the pressure on the three remnant fens from more intensive agriculture, and to expand the habitats.

The relationship between the Anglesey Fens and those of eastern England goes beyond their wildlife value. The ancient physical connection of a royal road was used by Brittonic warrior Queen Boudicca to traverse Britain from her Iceni tribe homelands in East Anglia to North Wales and across the sacred sea-crossing to the Druidic stronghold of Mona. The considerable wealth of the Iceni people was held on the island until the Roman defeat of the Druids on Anglesey, and the later defeat of Boudicca and her armies following a revolt, possibly arising from the desecration of their religious centre. A Roman road then connected the two areas of land, allowing produce from the fertile arable land on Anglesey to be transported to other Roman centres.

Travel Notes

The railway station at Bangor has regular train connections to Anglesey across the Menai Strait by the Britannia rail bridge, built by Robert Stephenson in 1850, to Llanfairpwll station (a request stop) that serves the village of Llanfair Pwllgwyngyll. The village is more famous for its longer, albeit synthetic name, Llanfairpwllgwyngyllgogerychwyrndrobwllllantysiliogogogoch.

National Cycle Route 5 follows quiet roads from Bangor across the Menai Bridge, built by Thomas Telford as the first ever major suspension bridge, to Pentraeth for Cors Bodeilio Nature Reserve and the village of Capel Coch for Cors Erddreiniog and Cors Goch reserves.

Cors Erddreiniog National Nature Reserve is located between Brynteg and Capel Coch. Here there is an access point (SH 458 822) for a long track heading east onto the fen with a boardwalk through the reed and sedge-beds towards Llyn yr Wyth Eidion, the lake of the eight oxen, which legend has it fell into the peaty pool along with the unfortunate ploughman.

Cors Goch National Nature Reserve is owned and managed by the North Wales Wildlife Trust. The entrance is situated west of Pentraeth. From the village head north along the A5025 for a short distance to a junction and turn left towards Llanbedrgoch. After 0.5 kilometres there is a signpost for the reserve and a parking area (SH 504 816) with a track leading to the reserve.

Cors Bodeilio National Nature Reserve is managed by Natural Resources Wales. From the main street (A5025) in Pentraeth, head west just after the pelican crossing, signposted to Llangefni, then left again towards the school. Continue down the narrow winding lane for 2.4 kilometres. The entrance to the reserve and a small parking space are on the right (SH 506 772). There are marked trails and a boardwalk.

Place names

Llanfair Pwllgwyngyll: the church of St Mary (*Llanfair*) by the pool (*pwll*) of the white hazels (*gwyn gyll*).

Llangefni: parish of the River Cefni, derived from the Welsh *cafn*, meaning 'dip', 'hollow' or 'trough'. This probably refers to the narrow gorge now called Nant y Pandy or The Dingle, on the outskirts of Llangefni.

Cors Erddreiniog: most likely is a mixing of Welsh and English over time from *cors wedi ddraenio* meaning 'the marsh that has been drained'.

Cors Goch: Welsh, meaning 'red marsh'.

LAKE VYRNWY

Map: OS 1:50,000 Sheet No 125
Peatland Grid ref: SJ 016 192
Access point: RSPB Centre (SJ 016 192), Rhiwargor (SH 965 241)

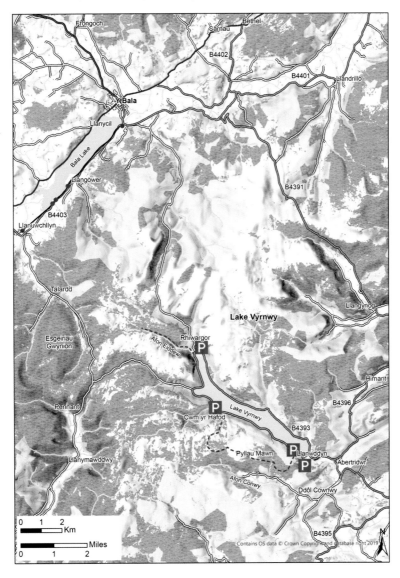

Lake Vyrnwy nestles in the Berwyn mountain range in Powys to the south-east of the Snowdonia National Park. The setting is a remote upland area with a picturesque mosaic

Top of Rhiwargor Falls, Lake Vyrnwy

Bryn Mawr, bog pools, Lake Vyrnwy

of farm pastures and lakeside oak woodlands giving way to heather moorland and blanket bog on the higher ground. The surrounding hills and mountains are designated as the Berwyn National Nature Reserve covering some 8,000 hectares and the most important upland area in Wales for breeding birds, including hen harrier, golden plover, curlew and red kite (*Milvus milvus*). There is also a chance to see black grouse near the southern extremity of its range in Wales. One of the features of the blanket bogs here are the occasional patches of Cloudberries (*Rubus chamaemorus*), an alpine plant that resembles low-growing blackberry plants except that the berries are a golden colour. The fruit is rather bitter with many pips but is much favoured by birds and a popular delicacy in Scandinavian countries, where they are made into jam.

The lake was formed by the construction of a stone dam in the 1880s, the first of its kind and which created the largest artificial reservoir in Europe at the time. The old village of Llanwddyn and its church were demolished to make way for the reservoir, which was constructed to supply drinking water piped to the city of Liverpool over a hundred kilometres away. The village derives its name from Saint Wddyn, a hermit who according to legend survived only on wine and would feed all his meat to two ravens, Huginn and Muninn, who brought him worldly knowledge in return. During periods of low water the eerie structures of the submerged ruins of Llanwddyn pierce the lake surface.

Before it was flooded, the valley was known as a boggy location, much of it underwater over winter with large areas of rushes, alder and willow. Farmers in the 19th century would cut peat from the moors as fuel to heat their cottages. More recently the moors were drained and burned frequently for agriculture, with forestry conifer planting across the peatlands.

The lake and much of the surrounding hills are now owned by the Hafren Dyfrdwy (Severn Dee) water company and include the RSPB Cymru Lake Vyrnwy Nature Reserve. The land is managed by RSPB Cymru as an upland organic farm with sheep grazing, and it has a wealth of peatland birds such as golden plover, hen harrier and merlin, with a few curlew on the lower ground among a wide range of woodland and farmland species. The moors are no longer burned and large areas of the damaged peatland have been repaired by blocking drains and removing conifers, benefitting both the peatland habitat and the water quality in the reservoir. The site is a showcase example of how farming in such areas can provide quality livestock as well as other products of benefit to society, including enjoyable wildlife-rich countryside, natural carbon stores and drinking water improvements.

Travel Notes

The RSPB Cymru Lake Vyrnwy reserve has a shop and visitor centre at the end of the dam in the village of Llanwddyn. The nearest rail station is Welshpool, thirty-seven kilometres away. Regular buses run from Welshpool to Llanfyllin Market Square, twenty-four kilometres from the reserve, but the connection to Llanwddyn operates only a few times a week.

The long-distance national trail, Glyndŵr's Way, connects Lake Vyrnwy and Welshpool using quiet country lanes and tracks. The trail is named after Owain Glyndŵr, the medieval Welsh prince who organised a rebellion against the English King Henry IV in 1400.

Information boards and leaflets at the RSPB Cymru shop offer several waymarked routes leading from the lake. For the best mountain views, follow the Llechwedd-du Trail that starts opposite the boat house car park (SJ 014 194) on the B4393, half a kilometre north-west of Llanwyddn village. The route involves a gentle climb through the forest before reaching the open heath and bog of Pyllau Mawn, and then descending into Cwm yr Hafod.

At the north end of Lake Vyrnwy, the trail from the car park (SH 963 240) and bird hides at Rhiwargor leads through forest in the valley of the Afon Eiddew, to the spectacular cascades of the Rhiwargor waterfall, known locally as Pistyll Rhyd-y-meinciau. From here there are excellent views of the surrounding peatlands of the Berwyns.

Place names

Berwyn: the white summit from the Welsh *barr*, meaning 'top' or 'summit', and *gwyn*, meaning 'white' or 'fair'.

Rhiwargor: 'slope of the defensive enclosure', from the Welsh *rhiw*, meaning 'brow' or 'slope' and *argor*, 'a palisade'.

Pistyll Rhyd-y-meincau: Welsh, meaning 'waterfall of the ford of benches/cascades'.

Lake Vyrnwy: from the river Efyrnwy, derived from *maranwy*, meaning 'salmon river'.

Llanwddyn: Welsh, 'parish of the hermit named Wyddn'.

MID WALES

The ancient county of Ceredigion stretches from the high summits of the Cambrian Mountains westwards across moorland and pastures down to the coast between Cardigan and Aberystwyth. Ceredigion has long been known for its dairy farming and was once the main supplier of milk to London. Above the lush pastures, the highest point in the Cambrian Mountains is Pumlumon, or in its anglicised form Plynlimon, a dominating massif whose blanket bog contains the source of the River Severn, Britain's longest river. The Montgomeryshire Wildlife Trust has a large nature reserve at Glaslyn (SN 826 941) with blanket bog and heathland habitats. They have embarked on a huge landscape-scale project, centred around Pumlumon and covering 40,000 hectares of upland habitats between Aberystwyth and Llanidloes. This is part of the 'Summit to Sea' initiative where environmental bodies, landowners and businesses are working to conserve and restore habitats from the Cambrian Mountains to the Dyfi estuary.

To the south of Pumlumon, the Cambrian Mountains contain a vast stretch of moorland, in an area called Elenydd surrounding the Elan River and its series of large reservoirs. Sometimes referred to as 'The Desert of Wales', centuries of intense livestock grazing and burning management have left the peatland eroding and the surface damaged to such a degree that the area is dominated by purple moor grass that is of little value to wildlife or grazing animals, nor indeed to carbon sequestration and storage. Parts of the area are owned by Welsh Water and managed by the Elan Valley Trust. A visitor centre just outside Elan Village provides information about plans to restore the peatlands.

Much of Elenydd once belonged to Cistercian monks based at the Strata Florida Abbey. During medieval times the monks had considerable influence on the upland landscape, managing woodlands for charcoal and cutting peat from hillslopes and the nearby lowland raised bogs of Cors Caron, beside Tregaron in the floodplain of the River Teifi. Another of Britain's largest and best raised bogs, Cors Fochno sits on the southern shores of the Dyfi estuary.

South of the Elan Valley, Abergwesyn Commons, managed by the National Trust, is an upland area containing large areas of blanket bog on the southern edge of the Cambrian Mountains between the Nant Irfon valley and Llanwrthwl, with long-distance walking routes across the area.

The region is regarded as the heartland for red kites, a bird of prey whose last stronghold in Britain consisted of just a few pairs in this part of Wales at the start of the 20th century, following persecution and habitat loss. From here conservation work has seen a massive recovery, with a thriving population across Britain. The birds can regularly be seen hunting over the raised bogs.

Strata Florida Abbey ruins, Pontrhydfendigaid, Ceredigion

CORS FOCHNO

Map: OS 1:50,000 Sheet No 135
Peatland Grid ref: SN 627 911
Access point: Cors Fochno Trail car park (SN 633 921)

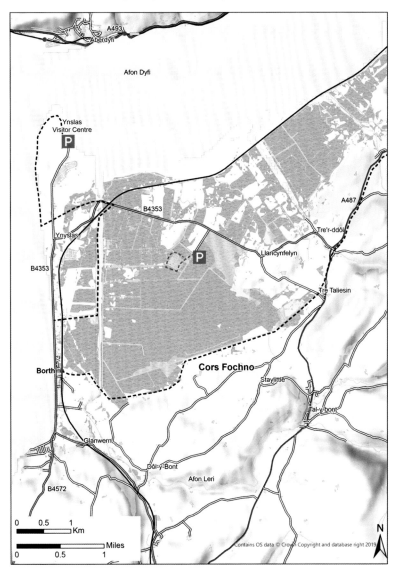

Cors Fochno, also known as Borth Bog, is a large estuarine raised bog in the north of Ceredigion adjacent to the popular coastal holiday village of Borth. Despite a long history of peat cutting and agricultural drainage around the margins, the remaining central dome of peat is one of the largest estuarine raised bogs in Britain. The name of the peatland Cors Fochno is traditionally thought to mean 'bog of the pigs', so named by the locals after a prince exploring his realm brought the first domestic pigs to the area.

Now part of the Dyfi National Nature Reserve and a UNESCO Biosphere Reserve, the bog is mostly owned and managed by Natural Resources Wales. Occupying the beach between Borth and Ynyslas is a 'sunken forest', an incredible scene of ancient tree stumps revealed in the sand at low tide and in extreme weather. These are the preserved remains of a forest of pine, alder, oak and birch that grew here around 5,000 years ago, before the peatland formed, and which was then eroded by the estuary waters.

The area's archaeological importance also includes discoveries of a major Iron Age and Roman lead smelting site on the bog margins south of Llangynfelyn, near the 19th-century lead mines at Taliesin, also known as *Pwll Roman*, meaning 'the Roman pits (mines)'. Excavations of a medieval wooden trackway around the smelting works revealed that after they had been abandoned by the Romans the area was inundated by peatland, and centuries later in the mid-10th century the trackway was constructed over the bog to reach the higher ground of Llangynfelyn.

The bog remained in low-impact use until the late 18th century, when agricultural intensification led to drainage ditches and the River Leri being diverted and canalised along a route closer to the estuary. By the 19th century almost two-thirds of the original bog area had been turned into pasture and arable land. In 1863 a railway was constructed across the bog and a new road, the current B4353, and farmsteads were developed along the north of the bog. The increased access allowed peat to be removed for fuel, with farmers sharing a steam-driven turf cutter.

Burning has also been a major problem at Cors Fochno, with fire being used to improve livestock grazing but sometimes getting out of control and damaging large parts of the bog. Despite the site being given protected status in 1954, drainage and fires continued until it was purchased by the then Nature Conservancy Council in 1981. Occasional fires are still a problem because the bog remains drier than it would be had it not suffered such major drainage impact. A large fire was started during a storm in 2014, when sparks from electricity poles were blown by the wind.

The Cors Fochno toad is mentioned in a Welsh legend about the oldest animals. Written accounts from the 16th century by Thomas Williams of Trefriw tell of the toad who when

Drawing: White-fronted geese

asked about his age replied, 'I've never eaten anything but what I could get from the earth, and I never got enough of that. Do you see the two large hills by the bog? I saw that land when it was flat. And nothing has made them except the little I have excreted.'

From the mid-20th century, the area's wildlife drew attention from scientists at Aberystwyth University, leading to its protection. Rare insects such as bog bush cricket (*Metrioptera brachyptera*) and rosy marsh moth (*Eugraphe subrosea*) have been found here, more than a century after disappearing from the fens of eastern England. The bog and surrounding wet pasture support breeding waders, with the reedbeds supporting grasshopper warblers, reed warblers and sedge warblers. In winter the bleakness can be brightened by birds of prey, such as short-eared owl and occasionally hen harrier. The centre of the bog has the best representation of sphagnum mosses of any southern Britain peatland and is an important site for the insect-eating great sundew. Royal fern (*Osmunda regalis*) can be found growing along the ditch margins.

In 2014 a programme of conservation restoration saw an impressive sixty kilometres of low peat banks, known as 'bunds', being constructed to help maintain water levels at a height that can allow sphagnum mosses and other peatland vegetation to thrive. This work has since been extended as part of an EU LIFE project.

Travel Notes

The railway station at Borth is on the line between Machynlleth and Aberystwyth, with good views of the bog from the train. A bus service from Aberystwyth to Tre'r-ddol goes via Borth and Ynyslas, where the Ynyslas Visitor Centre for the nature reserve is located.

There are two long-distance walking routes that pass Cors Fochno: the Ceredigion Coast Path from Borth along the west side of the reserve to Ynyslas, and the Wales Coast Path from Borth to Glandyfi around the bog's southern edge. Both these routes go through wet pastures, reedbeds and scrub, often on old peat cuttings around the bog edge. The Cors Fochno Trail goes onto the bog itself and is reached by a track on the south side of the B4353, three kilometres from Tre'r-ddol. A gate at the start of the track (SN 636 926) leads to a small parking area and access to a circular boardwalk around the bog.

Place Names

Dyfi: Welsh, probably meaning 'water'.

Ynyslas: Welsh, meaning 'blue island'.

Llangynfelyn: Welsh, meaning 'the parish dedicated to the church of St Cynfelyn'.

CORS CARON

Map: OS 1:50,000 Sheet Nos 135, 146
Peatland Grid ref: SN 690 640
Access point: Tregaron car park (SN 692 625)

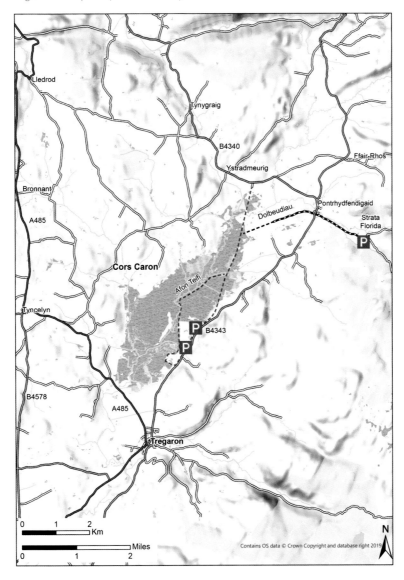

Set within picturesque pastures and woodland nestled below the Cambrian Mountains, Cors Caron is a large area of raised bog filling the high-level valley of the upper River

Willow arch entrance to Cors Caron

Cors Caron

Teifi between Tregaron and Pontrhydfendigaid. Extending over 800 hectares, the site is comprised of three raised bog domes formed over 12,000 years ago after glacial action left a large shallow lake, later draining naturally to be replaced by oak woodland and pools that filled with peat in the area surrounding the meandering River Teifi. The area has also been known as Tregaron Bog, from the nearby village of the same name, and translates as 'the homestead of Caron' (a lady's name). Early maps from the late 19th century also give the name *Cors Goch Glanteifi*, meaning 'red bog on the banks of the Teifi'. This is one of Britain's best and largest raised bog sites, famous for its spectacular panoramic views and surrounding mix of habitats with reedbeds, wet grasslands, woodlands and rivers. It is likely that this peatland landscape consisted originally of five or more discrete but interconnected raised bog domes.

Despite their idyllic setting, the peatlands were once a hub for local economic activity and have been greatly altered over the centuries. The 12th-century Cistercian monks at Strata Florida Abbey, a few kilometres to the north-east of Cors Caron, were skilled at mining the rich ores of lead and silver in the vicinity of Pontrhydfendigaid, a market town established to house the mine workers. Peat was extracted from Cors Caron to fuel the smelting process and resulted in the destruction of former peat bogs, now agricultural land to the extreme south and north of the current site. Tregaron was partly built on former peatland as one of the granges for Strata Florida. After the dissolution of the monasteries, peat continued to be cut by hand as fuel for the surrounding farmsteads. In the early 20th century peat was extracted mechanically for animal bedding but by the 1960s all peat cutting had stopped.

In the 1860s the Manchester and Milford Railway Company constructed a line between Pencader Junction and Aberystwyth that crossed the eastern side of the bog rather than cut through the adjacent, more 'valuable', farmland. Victorian engineer David Davies devised the idea of using wooden faggots and bales of wool to provide a floating base for the railway and bought up all the local wool supplies in the process. The results had mixed success, as sections of track sank and required ongoing maintenance until the line's closure in 1965.

Today Cors Caron is owned and managed by Natural Resources Wales, which has constructed trails along the disused railway and boardwalks across wetter terrain. A major EU-funded restoration project has been underway here and on other outstanding Welsh bogs since 2018. The large area and diverse mix of habitats make for a thrilling wildlife experience. The large heath butterfly is found here, and there are over sixteen species of dragonfly, including black darter and common hawker. One of the very special insects on Cors Caron is the rosy marsh moth, thought to have become extinct in Britain over a hundred years ago but discovered here in recent years.

Snipe, curlews and water rail (*Rallus aquaticus*) breed here, and in winter the site supports hen harrier and red kites and is important for wildfowl, particularly whooper swans, with occasional occurrences of wood sandpipers (*Tringa glareola*) and green sandpipers (*Tringa ochropus*).

Travel Notes

Buses run from Aberystwyth rail station to Tregaron. The main entrance to Cors Caron is situated three kilometres away from the village, directly off the B4343 on the left-hand side (SN 692 625).

The thirty-two-kilometre National Cycle Route 82, the 'Ystwyth Trail', mostly uses disused railway lines to connect Aberystwyth and Tregaron. A 3.5-kilometre boardwalk leads through the reserve to bird hides and there is a trail along the disused railway. The riverside trail can be dangerous and difficult to pass in winter.

The ruins of Strata Florida can be reached by leaving the railway track at Dolbeudiau farm (SN 699 664). It was in this area in 1811 that a preserved headless bog body, possibly from an Iron Age ritual burial, was recovered by peat cutters and subsequently reburied in Ystrad Meurig churchyard.

From the farm, head east to Pontrhydfendigaid and then two kilometres south-east along the lane to the abbey. There is a marked track known as the Monks' Trod, an ancient by-way originally connecting the Abbeys of Strata Florida and Cwm-Hir near Rhayader. The route passes through the blanket bogs of the Elan Valley to Pont ar Elan at Craig Goch Dam Reservoir.

Place Names

Tregaron: from the Welsh *tre*, meaning 'town', 'hamlet' or 'homestead', and *Caron*, 'a person's name'.

Pontrhydfendigaid: from the Welsh for 'bridge at the holy ford'.

Strata Florida: from Latin, meaning 'vale of the flowers'.

Dolbeudiau farm: from Welsh *dol* ('low-lying point') and *beudy*, meaning cow house.

SOUTH WALES

South Wales is best known for its coal mining heritage, with large coalfields south of the Brecon Beacons mountain range between Merthyr Tydfil and Swansea. Formed about 300 million years ago, the coal is the preserved and compressed remains of tropical peatland formed from forests of tree-like giant club mosses, *Lycophytes*, that could grow up to fifty metres tall.

The main areas of blanket bog in the region are in the Brecon Beacons National Park, a mountainous area of old red sandstone peaks, to the south of Brecon. However, this iconic British peatland type extends further south into the heads of the South Wales valleys and even further south again into the lowlands in places. Intensive livestock grazing and burning have left much of the peatland damaged and eroding. There are several restoration projects underway within the park, including large areas of the Black Mountains, the easternmost range of the Brecon Beacons. In other parts of South Wales, upland areas that were extensively afforested with conifers since the 1950s are being restored to peatland. In an area blighted by forest fires on the drained and dry peat, the 'Lost Peatlands of South Wales' is a re-wetting project in what used to be referred to as the Alps of Glamorgan, between Neath Port Talbot and Rhondda Cynon Taf.

In the lower-lying lands there are many remnants of fen peatland and lowland raised bog within the agricultural landscape. Crymlyn Bog near Swansea is a fen and one of the largest examples of the habitat in Wales. The Wildlife Trusts of South Wales have several raised bogs nature reserves, the largest of which is Cors Goch, Llanllwch, near Carmarthen.

Ogof Ffynnon Du, Brecon Beacons

CORS CRYMLYN

Map: OS 1:50,000 Sheet Nos 159, 170
Peatland Grid ref: SS 695 945
Access point: Cors Crymlyn Visitor Centre (SS 685 942), Pant y Sais car park (SS 715 944)

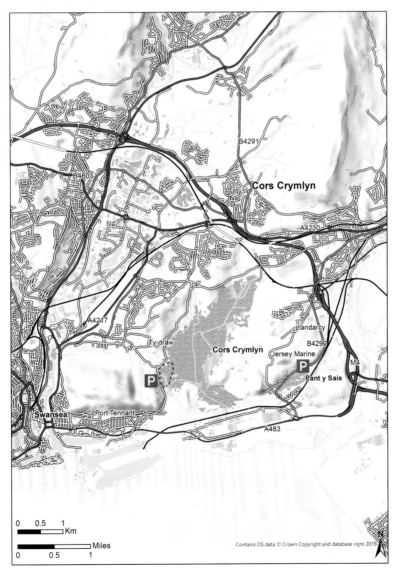

Cors Crymlyn is an oasis of green to the east of Kilvey Hill, in the heavily populated industrial landscape around Swansea on land that was once the floodplain for the River

Pant-y-Sais boardwalk

Lily pond, Cors Crymlyn

Clydach and River Neath. Formed in a large depression, hollowed out when glaciers retreated at the end of the last Ice Age, Crymlyn Bog as it is known in English is actually a fen and the largest example in Wales. Similar to the fenlands of eastern England, it contains tall reedbeds and sedges surrounded by waterlogged willow scrub. Among the varied fen peatland there are also areas of more acid sphagnum bog forming part of the transition mire and quaking bog habitat. Designated as a National Nature Reserve, the site includes a smaller fen a few kilometres to the east at Pant y Sais.

With heavy industry in close proximity, Cors Crymlyn has suffered from oil spills from a nearby oil refinery and was used as a dumping ground for ash from coal-fired power stations as well as being a municipal landfill site. In the late 1970s, the plight of Cors Crymlyn came to the attention of Andrew Lees while he was working with the Nature Conservancy Council. With his experience growing up in the Norfolk fens he had a great affinity with the site and helped prevent it from being used for rubbish tipping. Later, working with Friends of the Earth, he campaigned for the site to be protected under conservation law and eventually won the battle, with the site being declared nationally and internationally important in 1993. Sadly, Andrew died a year later at the age of forty-six in a

Drawing: Adder

Madagascar forest, while filming for a campaign there. Andrew was recognised as one of the great environmentalists of his time and a memorial was constructed at Pant y Sais fen in his honour. The inscription describes him as 'the man who started the fight to get the bog protected' and carries a quotation from him: 'At some point I had to stand up and be counted. Who speaks for the butterflies?'

The wildlife of the area is exceptional. It is home to one of the few populations of fen raft spider in the UK. Another rare species, royal fern, so called because of its grand size, is found particularly on Pant y Sais along with yellow iris (*Iris pseudacorus*), bog bean and a host of colourful flowering plants. The reserve is a haven for the many reedbed bird species, including water rail and bearded tit, and occasionally a bittern's booming call can be heard.

Travel Notes

There is a railway station at Swansea with a regular bus service to Port Tennant, about one kilometre from Cors Crymlyn. Entry to the main car park (SS 685 942), where the visitor centre is located, is via Dinam Road just north of the Fabian Way park-and-ride in Port Tennant. The reserve has several trails and a boardwalk.

The Pant y Sais car park (SS 715 944) is directly off the B4290, east of Jersey Marine village.

There is a bus stop opposite the entrance to the reserve boardwalk, which has a regular service from Swansea and Neath. The Tennant Canal towpath between Swansea and Neath also passes the reserve.

Place names

Crymlyn: from Welsh *crymu*, meaning 'to bow or curve' and *llyn*, meaning 'lake'.

Wern Road: from Welsh *gwern*, meaning 'wet, boggy area' or 'swamp'.

Pant y Sais: from Welsh for 'hollow or valley' (*pant*) and 'Englishman' (*Sais*).

River Clydach: Welsh *clydach*, from Brittonic origin *klou*, meaning 'strong flowing', 'swift' or 'stony'.

IRELAND

Ireland's westerly location, strongly influenced by an Atlantic climate with moisture-laden clouds and high rainfall, is ideal for peatland plants and peat formation. There is an east-to-west gradient in the climate, giving a diverse range of peatland types. Raised bogs and fens are largely associated with the low-lying midland plains with swathes of blanket bog adorning the west coast mountains, hills and glens. Peatlands originally covered twenty per cent of the island of Ireland, but over seventy-five per cent of that area has lost its conservation value.

For many people, peatlands are characteristic of Ireland. Gardeners across Britain have literally been taking home a piece of Ireland, at the rate of millions of tonnes each year, in the compost and growbags that adorn our garden centre shelves. Popular images in print and film portray idyllic scenes of humble cottage-dwelling turf cutters. The Irish Gaelic language gives us the term 'bog' from *bogach*, meaning 'soft and marshy ground'. Those who have lived and worked among the peatlands have often regarded them as inconvenient, without value, and in need of 'improvement'. Not surprisingly, agricultural intensification in the early 20th century was viewed as opportunity, and much of the peatland was drained and fertilised.

Around that time, poets and artists began capturing the new identity of Ireland's traditional, rural way of life, stirring a latent passion and sadness at what was being lost. The drive for economic improvement prevailed, however, and governments began investing in programmes of peatland exploitation, particularly in the raised bogs of the Irish midlands, with commercial extraction to fuel electric power generation and to supply overseas horticulture demands. In the 1980s agricultural subsidies saw increased peatland drainage and a threefold increase in sheep, whose grazing led to vast areas of peatland erosion, particularly in the uplands. More recently, forestry expansion and windfarms have seen large areas of blanket bog in particular being damaged and, in extreme cases, leading to dramatic and devastating bog bursts where tonnes of peat and water end up sliding down hillsides.

A strong peatland conservation movement began in the late 20th century in Ireland. Aided by European Union intervention, much of the domestic-scale peatland damage on nature conservation sites was halted, in some parts because the peatland resource had been exhausted, and a new era of conserving and restoring the least damaged peatlands began. Domestic peat cutting for fuel is still visible on raised bogs across the midlands and the blanket bogs of the west coast, but also apparent is the huge public interest in the wildlife and heritage of the peatlands. Ireland still supports half of all the raised bogs remaining in North-West Europe and eight per cent of the global blanket bog distribution.

Government funding has gone into developing national parks with visitor centres and interpretation facilities in some of the large peatlands. Community groups have become involved in looking after and repairing the bogs on their own doorstep. The surging demand for access into the countryside has brought its problems and managing the movement of so many visitors is being addressed in part by the provision of wooden boardwalks across sensitive peatland areas. These imposing features on the peatland landscape are not so alien. For thousands of years, Ireland's post-glacial inhabitants revered the peatlands as sacred rejuvenating places, and wooden trackways onto the bog provided routes for religious ceremonies in which offerings, including human sacrifices, were placed in the peat and were preserved until they were exposed by peat cutters or erosion. Many of these fascinating artefacts, including bog bodies, can be seen in the National Museum of Ireland in Dublin.

Snipe over blanket bog, Ireland

NORTHERN IRELAND

Blanket bogs occupy the mountains and hills that surround Northern Ireland, with the Sperrin Mountains in the north-west, the Antrim Plateau in the east and the Cuilcagh Mountains bordering with the Republic of Ireland in the south-west. Raised bogs are mostly found in the centre around the Lough Neagh basin and north along the Bann river valley. These peatlands once covered eighteen per cent of the land area but less than fifteen per cent of the blanket bog and ten per cent of the raised bog habitat now remains.

Although Northern Ireland's peatlands largely escaped government-sponsored industrial peat extraction, they were widely cut for domestic use and huge areas were lost to agricultural and forestry expansion in the 20th century. Intensive farming, particularly the poultry industry, has also resulted in high levels of nitrogen pollution that is having a damaging impact on the sensitive peatland vegetation, including the mosses and insect-eating plants that are naturally suited to a low-nutrient environment.

The capital city of Belfast offers an immediate upland, blanket bog experience in the surrounding Belfast hills. Black Mountain and Divis (from the Irish *dubhais* meaning 'black back') are two mountain peaks within the Belfast hills. There are several trails offering excellent views over the city as well as Lough Neagh. Visitors are asked not to go over the blanket bog itself but to keep to the summit ridge and lough trails to avoid further damage to the delicate bog vegetation. Further details of the routes are available from the National Trust.

Northern Ireland Water has been working with environmental organisations on restoring a number of blanket bogs for conservation and to improve drinking water quality. Work includes blocking old drainage ditches in Northern Ireland's largest single area of blanket bog: the Garron Plateau above Glenariff in the Antrim hills. Other projects include removing conifer plantations from an area of blanket bog on the Pettigoe Plateau, north-west of Lower Lough Erne, County Fermanagh.

One of the most famous blanket bog areas in Northern Ireland is on Cuilcagh Mountain, straddling the border between County Fermanagh and County Cavan in the Republic of Ireland. The lower-lying land around Lough Neagh supports most of the remaining raised bogs in Northern Ireland. One of the largest remnants is the Ulster Wildlife Trust Reserve at Ballynahone Bog near Maghera in County Derry. This was the site of public protest against government-sponsored peat extraction for horticulture. At the start of 1991, near-intact bog was drained as preparation for peat extraction. A few years later the development was halted and the company responsible blocked the drains. Visitor access is limited to allow the site to recover. The Peatlands Park in County Armagh is a former peat cutting site on a raised bog where there are excellent visitor facilities explaining the area's natural and cultural history.

Boardwalk, Cuilcagh Mountain

PEATLANDS PARK

Map: OSNI 1:50,000 Sheet No 19
Peatland Grid ref: H 897 610
Access point: Peatlands Park car park (H 899 602)

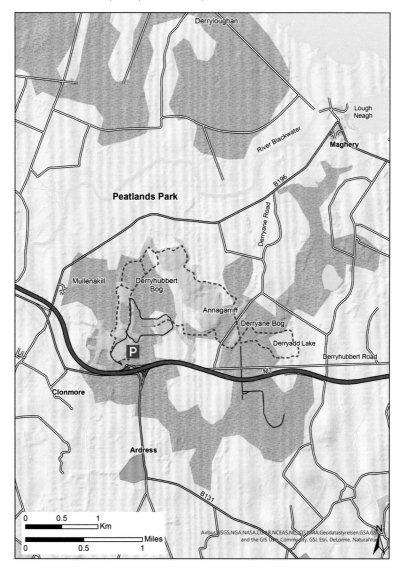

The Peatlands Park is located in County Armagh near Dungannon, by the southern shores of Lough Neagh. The area contains two National Nature Reserves: Mullenakill, an uncut

Carr woodland, Annagarriff, Peatlands Park

raised bog, and Annagarriff, an ancient woodland fringed by fen and raised bog. Between them lies Derryhubbert Bog, an area of cut-over raised bog. Formerly a private estate belonging to the Verner family for over four hundred years, the land was sold to the Irish Peat Development Company in 1900. The Northern Ireland Environment Agency now owns and manages the site as a country park, which opened in 1990.

The early 20th-century peat works included a narrow-gauge railway to cart the turves for processing and for export to England, mostly as animal bedding and for packing vegetables. Initially the carts were pulled by men, then donkeys, before electric and later diesel locomotives were deployed. During the First World War sphagnum moss was collected from the site for use by soldiers as wound dressings. Peat cutting was replaced by milling to produce peat for horticulture, but all activity stopped in 1960, by which time three metres of bog surface had been removed. Subsequent subsidence and drying out of the peat left the area vulnerable to frequent fires. Once the site came under government ownership, drainage ditches were blocked to raise water levels and invasive rhododendron plants were removed. A large network of trails and boardwalks was constructed, with visitor interpretation explaining both the natural history of the site and the commercial peat cutting past.

The park offers a unique opportunity to see some of the finest examples of raised bog habitat with all its peatland species, as well as a rare example of ancient woodland at Annagarriff, protected as a hunting preserve for over two hundred years. Oak and birch

Former cut-over peatland, Derryhubbert Bog

Peatland railway, once used to carry cut peat, Peatlands Park

predominate along with yew, holly and aspen. The site also holds over ninety-five per cent of Northern Ireland's alder buckthorn (*Frangula alnus*), a rare thornless tree found in wetland areas. Fen and sedge surround the woodlands in a natural association no longer seen in many of our lowland peatlands. Despite the park's popularity, there is enough space for quiet appreciation of this wonderful diverse suite of habitats and its wildlife.

Travel Notes

The Peatlands Park is ten kilometres north-west of Portadown, with a railway station on the line from Belfast. There is a regular bus service to the Peatlands Park from here. National Cycle Route 94 goes from Portadown to the village of Maghery. From here the Birches and Maghery Cycle Trail uses quiet roads heading south-west out of the village on the B196 and then first left south on the Derryane Road to the Peatlands Park.

In the park there are several well-marked trails and a visitor centre with car parking accessed from Derryhubbert Road, north of the Loughgall exit on the M1 motorway.

Place names

Annagarriff: meaning 'rough bog'.

Mullenakill: meaning 'church on the hill'.

Derryane: from Irish *Doire Eithne*, meaning 'wood of Saint Eithne'.

Maghery: meaning 'a level plain or field'.

Dungannon: from Irish *Dún Geanainn*, meaning 'Geanann's fort'.

Derryhubbert: Irish, meaning 'Hubbert's wood'.

CUILCAGH MOUNTAIN

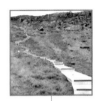

Map: OSNI 1:50,000 Sheet No 26
Peatland Grid ref: H 114 288
Access point: Marble Arch Visitor Centre (H 122 344)

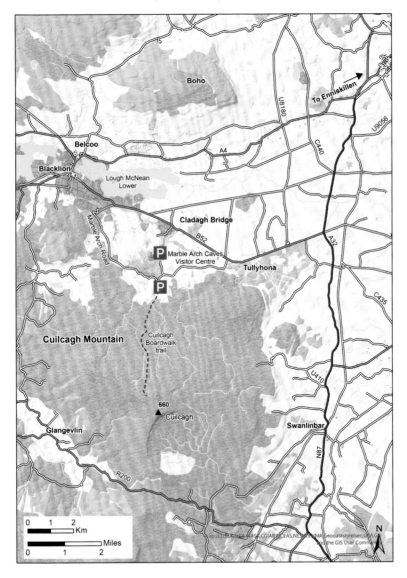

Cuilcagh Mountain is an imposing dominant feature rising up from the broad plains of the Lower Macnean Valley in County Fermanagh and visible from much of the northern

Boardwalk and steps 'stairway to heaven', Cuilcagh Mountain

Limestone outcrop, among blanket bog, Cuilcagh Mountain

half of Ireland. Straddling the Border with County Cavan in the Republic of Ireland, the region is often referred to as the Cuilcagh Mountains with the name Cuilcagh coming from the Irish *Binn Chuilceach*, meaning 'chalky peak'. The summit is actually millstone grit on top of shale and siltstone, with the lower slopes composed of limestone formed over 300 million years ago when the land that is now Ireland was covered by a shallow tropical sea near the Equator. Water flowing off the mountain percolates through deep blanket bog, before entering a network of caverns forming the famous Marble Arch Caves. The mountain's northern slopes and caves were included in the UK's first European designated geopark and were later included in the global geopark network.

This is one of the best areas for blanket bog in Northern Ireland. It is an important upland breeding site for golden plover but numbers have been declining in recent years. It is also a prominent feeding and roosting site for Greenland white-fronted geese. Since medieval times people had farmed the lower slopes and cut peat by hand for fuel. The Irish Famine in the mid-19th century forced people to leave the land, with only the ruins of stone farm cottages and field walls as a reminder of those lost communities. The area was drained and burned as grazing land for large numbers of sheep, and peat was extracted mechanically with tractor-drawn machines, forming huge 'sausages' of extracted peat for use as fuel.

By the 1980s, the bog had eroded to such an extent that rainwater began entering the upper caves at a faster rate. Severe flooding resulted in the caves below, threatening what was by then a major visitor attraction. In 1998, Fermanagh District Council together with the RSPB successfully secured funding from the European Union to establish the Cuilcagh Mountain Park and begin restoring the damaged bog. Twenty years later, the job of repairing eroded areas and blocking drains continues. The white fluffy heads of cotton-grass mark out where the old peat cuttings have been re-wetted as these are among the first peatland plants to respond to the restoration.

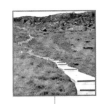

Cuilcagh Mountain has drawn people for thousands of years. The flat-topped summit and slopes are dotted with prehistoric hut sites, megalithic tombs, boulder monuments and cairns. The largest cairn on the plateau is the remains of a Bronze Age burial mound. Megalithic monuments located ten kilometres south-west of the summit at Cavan Burren in the Republic of Ireland are aligned with the Cuilcagh Mountain burial mound. This fascination with the mountain is not just an ancient phenomenon. The route to the 665-metre summit from a starting point near the Marble Arch Caves has attracted modern-day hikers – contemporary pilgrims seeking the elation and feeling of self-achievement that must have drawn our distant ancestors.

Since the origins of the Cuilcagh Mountain Park in 1998, attempts have been made to protect the delicate blanket bogs and rare alpine habitats of the mountain summit from the impact of visitor trampling causing erosion. In 2014 a controversial proposal for a huge wooden boardwalk across the bog and up the steep mountain side was approved, and construction began on a 1.6-kilometre structure with 450 wooden steps across thirty-six flights. Popularly called 'the Stairway to Heaven', and widely publicised on social media, the authorities were overwhelmed by the public response, as visitor numbers rose from 3,000 people a year to over 70,000, with a single Easter weekend exceeding previous numbers for a whole year. Disruption for local residents, car parking problems and increased pressure on the summit plateau habitats have led to temporary closures of the boardwalk while plans are developed to manage visitors through sensitive path work and education.

Travel Notes

Enniskillen is the nearest main town but has no railway station. There are frequent buses from there to the Marble Arch Caves Visitor Centre. Using public transport, walking or cycling are important in reducing pressure on car parking.

The Cuilcagh Boardwalk Trail is a 7.5-kilometre route (part of the long-distance Cuilcagh Way) that starts from a small car park on the Marlbank Road, 500 metres from the entrance to the Marble Arch Caves. Brown road signs for Cuilcagh Mountain Park give directions from the Marble Arch Road and there is an information panel in the car park with additional details of the trail. Staying on the marked routes is vital in reducing harm to the delicate alpine and peatland habitats.

Place Names

Enniskillen: meaning 'Ceithlenn's island', from the Irish *inis*.

Legnabrocky: from Irish *lag* ('hollow') of the badger.

REPUBLIC OF IRELAND

The west coast of Ireland experiences Atlantic weather conditions that are among the wettest in Europe, both in terms of the total rainfall (up to four metres a year in some mountains areas) and the number of rain days (up to 225 days a year). Far from being dreary, however, the dramatic mountains from County Mayo in the north through County Galway to County Kerry in the south hold some of the most important and spectacular blanket bogs in Europe. The Atlantic fringe is celebrated as a natural attraction, and the Wild Atlantic Way, one of the world's longest coastal routes, passes through a series of national parks showcasing peatlands and their wildlife.

The rural way of life in much of Ireland has left its mark. With woodlands long depleted, and limited coal deposits, peat was the country's alternative fuel. The intricate patterns of stacked turves set out to dry can still be seen across many peatlands. Over forty per cent of peatland loss has been due to centuries of hand cutting peat. The most dramatic exploitation came in the 20th century with the government-sponsored mechanised peat extraction, centrally controlled to provide fuel for electricity generation. The accessible and deep peat resource of the raised bogs occupying the midlands between County Offaly and County Kerry were systematically dug out leaving huge swathes of bare stripped peatlands, and in some areas exposing the underlying mineral rock and clay.

In a major turnaround, the peat company Bord na Móna set out a new strategy in 2016 to protect the remaining peatlands and invest in repairing damaged areas. Commercial peat extraction came to a formal end in January 2021, following an earlier High Court ruling requiring planning permission for such work. The company is now focusing its efforts on helping meet government climate change targets and restoring peatlands. The recovery process will take time. In the heavily damaged sites where little peat remains, fifty years of exploitation has set the clock back 10,000 years. The new low-carbon economy presents Ireland with opportunities for green jobs managing peatlands, income from tourism, and the development and supply of innovative peat-free products for horticulture.

The main national parks in the west of Ireland with good areas of blanket bog are Glenveagh in County Donegal, Ballycroy in County Mayo, Connemara in County Galway, and Killarney in County Kerry. In the east near Dublin, the Wicklow Mountains National Park stretches out through County Wicklow. Details of Killarney and Glenveagh are given in my previous book, *The Rainforests of Britain and Ireland*.

In the midlands, the Slieve Bloom Mountains are the stronghold for blanket bogs, but on the flat plains raised bog and small fens are the key features. The Irish Peatland Conservation Council, a wildlife charity, has been spearheading peatland conservation since 1982 and runs the Bog of Allen Nature Centre at Lullymore. The nearby Lullymore Heritage and

Discovery Park, geared as a recreational attraction for families, has information on the area's peat cutting history. Further north in County Longford, near Keenagh village, the Corlea Trackway Visitor Centre houses an excellently preserved Iron Age bog trackway made from oak planks over 2,100 years ago.

Oblong-leaved sundew, Roundstone Bog, Connemara

BALLYCROY

Map: OSI 1:50,000 Sheet Nos 22, 23
Peatland Grid ref: F 862 094
Access point: Ballycroy National Park Centre (F 804 099)

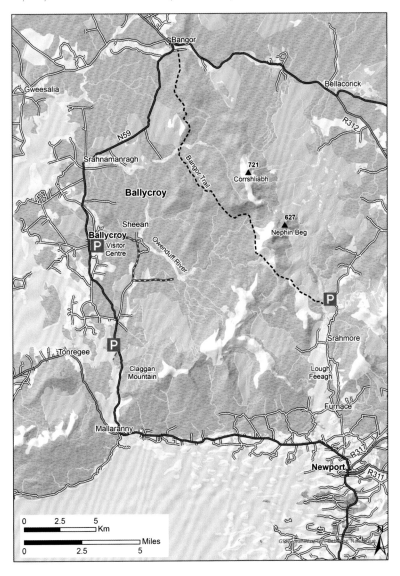

A huge expanse of blanket bog covering over 6,000 hectares around the Owenduff River spreads out from the coastal fishing village of Ballycroy beneath the Nephin Beg mountain

Claggan Mountain trail, Ballycroy

Pool with bog bean, Owenduff Bog, Ballycroy

range in County Mayo. Most of the area is within the Wild Nephin Ballycroy National Park. Owenduff Bog is internationally important as the best and largest remaining area of undisturbed blanket bog in Ireland. Commercial forestry planting, domestic peat cutting, burning and livestock grazing have damaged parts of the area, but its huge size allows the visitor to escape into as close an experience of wilderness as possible. The Irish naturalist Robert Lloyd Praeger described this landscape as 'the very loneliest place in the country' in his 1937 book, *The Way I Went*.

The pre-existing Ballycroy National Park was extended in 2017 to include the adjacent state-owned land of the Wild Nephin project, including 4,000 hectares of conifer forest. Under the original plan for the Wild Nephin, the commercial conifer forests were to be cleared, native woodlands allowed to grow and drained blanket bog re-wetted. There have also been discussions about reintroducing native red deer, which had become almost extinct in Ireland in the 20th century. They have a remaining stronghold in Kerry, though have not been in County Mayo for over 150 years. There is considerable debate about how much human intervention such wilderness plans should allow for. With so much pre-existing damage, including invasive rhododendron and alien conifers, it can be argued it would be irresponsible to simply let it go on without intervention.

There are a number of routes through the park, giving access to spectacular views of the peatlands. The Bangor Trail is an unofficial route along the old drovers' road used to take cattle from the north of the county to the coastal market town of Newport. The

Owenduff Bog can be viewed from minor roads leading east from the national park visitor centre at Ballycroy. There is also a boardwalk trail over blanket bog at Claggan Mountain to the south of Ballycroy. In an area known locally as Ros na Finne, 'the white wood of the waves', peat eroded by the tide has exposed ancient tree stumps, bleached by the elements, that grew in forests over 4,000 years ago, before the peatlands began to dominate.

On the north coast of Mayo near Ballycastle, a remarkable archaeological find in 1930 identified a Neolithic field system of stone walls under the peat. The area is called Céide Fields and has been dated back over 5,500 years. The people had arrived when the area was largely pine forest and had cleared the area of trees to rear cattle and provide arable land. Eventually the cleared land and a wetter climate led to increased peat growth and the community left with their traces preserved in the layers of peat that slowly accrued over the millennia. There is a fascinating visitor centre at Céide Fields and an opportunity to see the excavated stone walls.

Travel notes

A free shuttle bus operates during the summer months between Bangor and Newport, stopping at the Wild Nephin Ballycroy Park Centre and the Claggan Mountain Trail. There are also buses from Westport, where there is a railway station, and from Newport throughout the year.

Good views of the Owenduff Bogs can be seen from the minor road running east from Ballycroy towards the river at Shean. After three kilometres turn right at the crossroad and head south for 2.5 kilometres then turn left on a track heading east towards the river.

Céide Fields Visitor Centre (G 051 408) is open between June and September and is located beside the R314 road west of Ballycastle in County Mayo and can be reached by bus from Ballina to Balderg.

Place names

Céide Fields: *Achaidh Chéide*, meaning 'flat-topped hill fields'.

Owenduff: Irish *Abhainn Dubh*, meaning 'black river'.

Claggan: from Irish *An Cloigeann*, meaning 'a skull' or 'a round rocky hill'.

Nephin Beg: from Irish *Néifinn Bheag*; a small version of nearby mountain *Néifinn*, meaning 'sanctuary' or 'Finn's Heaven'.

Ballacroy: Irish *Baile Chruaich*, meaning 'town of the stacks', either hay or turf.

Mayo: Irish *Maigh Eo* meaning 'yew plain'.

CONNEMARA

Map: OSI 1:50,000 Sheet Nos 37, 44
Peatland Grid ref: Roundstone Bog L 682 446, Diamond Hill L 718 568
Access point: Derrigimlagh Bog car park (L 657 475), Connemara National Park Visitor Centre (L 711 573)

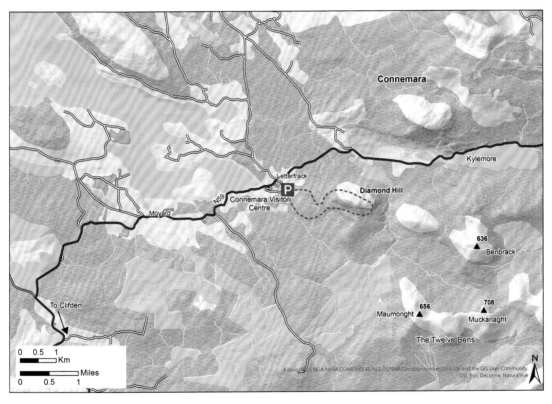

Connemara is a region of County Galway bounded to the west by the Atlantic Ocean. The Connemara Bog complex is an internationally important site containing large areas of blanket bog which thrives in the wet climate, even at low altitude, in contrast to other parts of Britain and Ireland where it generally occurs at higher levels. Such low-lying blanket bogs found in north-west Europe are often referred to as Atlantic blanket bog. The Connemara National Park contains a large expanse of blanket bog in the Diamond hills on the north side of the Twelve Bens mountain range. A smaller area of rocky lake-strewn blanket bog is situated further south, between Roundstone and Clifden.

Alcock and Brown memorial, Derrigimlagh Bog, Connemara

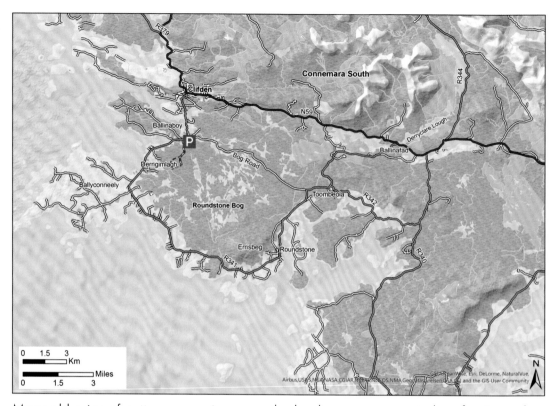

Many old ruins of cottages remain across the landscape as a reminder of a time when people managed to eke out a living before the country was drastically depopulated during the Great Famine in the 1840s. The main inhabitants now are sheep, whose grazing and trampling is a cause of conservation concern. The Roundstone Bog area includes Derrigimlagh Bog, containing the ruins of the Marconi wireless station where the first ever transatlantic telecommunication was broadcast in 1905. It is also the site where the first transatlantic pilots, John Alcock and Arthur Whitten Brown, crashed landed their biplane into the bog fourteen years later, after mistaking the flat bog surface for a firm grassy landing area. An interesting five-kilometre trail around a section of bog includes the Marconi building, with interactive displays along the route.

Errisbeg Hill stands out as a lone peak in the landscape and provides incredible views from the summit looking across the blanket bog studded by a multitude of small lakes. The winding bog road, laid out like an undulating ribbon across the peatlands from Ballinaboy Bridge, gives a real sense of being immersed in the wilderness. This road is feared by many local people who won't travel along it at night. A small pile of stones along the haunted bog road of Connemara is said to be the ruins of an 18th-century hostelry, called the Halfway House. The proprietors were a brother and sister who, legend has it, welcomed lone travellers then murdered them for their belongings and dumped the bodies in the peat.

Proposals for a Clifden airport runway at Ardagh, made in 1989, were opposed by local people concerned about the impact on a protected wildlife area. A later proposal at Derrigimlagh Bog met with objections at national level and was turned down by the authorities.

The Diamond Hill area within Connemara National Park has blanket bog and dry heath further up the hillsides. A rare member of the heather family, Saint Dabeoc's heath (*Daboecia cantabrica*), is found only in Connemara, South Mayo and parts of the Atlantic fringe of South-West Europe. Similarly, pale butterwort (*Pinguicula lusitanica*) and Saint Patrick's cabbage (*Saxifraga spathularis*) only occur in Ireland, Spain and Portugal, and not Britain. It is possible the plants arrived by sea as seed or were brought by ancient maritime traders. The popularity of this area increased dramatically with the opening of the national park and a long boardwalk had to be constructed to avoid the blanket bog habitats being eroded by so many visitors.

Travel notes

The nearest railway station is at Galway, over seventy-seven kilometres to the east of Clifden. Buses run from Galway and Clifden to the Connemara National Park Visitor Centre at Letterfrack on the N59. The main park entrance is two hundred metres south of the village, with marked trails and boardwalks round the peatlands on Diamond Hill.

For Roundstone Bog complex, leave Clifden on the R341 signposted to Roundstone. Turning left at Ballinaboy Bridge leads over the blanket bog via the bog road towards Toombeola Bridge. For an excellent circular cycling route, turn south two kilometres before the bridge onto the R341 to Roundstone and follow the road along the coast back to Ballinaboy Bridge. Five hundred metres before the bridge, a turning on the right leads down a track signposted to the Marconi building, the Alcock and Brown memorial, and the Derrigimlagh Bog Trail.

Place names

Connemara: derived from Irish the tribal name *Conmacne* and *mara*, meaning 'of the sea'.

Derrigimlagh: from Irish *Deirg-Imleach*, meaning 'red holm or strath'.

Errisbeg: from Irish *Iorros Beag*, meaning 'small border or peninsula'.

Toombeola: from Irish *Tuaim Beola*, meaning 'mound or burial place of Beola' (a mythical giant and leader of the *Firbolg* warrior tribe).

CENTRAL PLAIN

Map: OSI 1:50,000 Sheet Nos 48, 49, 55, 60
Peatland Grid ref: Bog of Allen N 696 254, Clara Bog N 249 303, Pollardstown Fen N 771 161, Abbeyleix Bog S 436 824
Access point: Bog of Allen Nature Centre (N 705 258), Clara Bog Visitor Centre (N 255 324), Pollardstown Fen car park (N 772 153), Abbeyleix Bog car park (S 435 840)

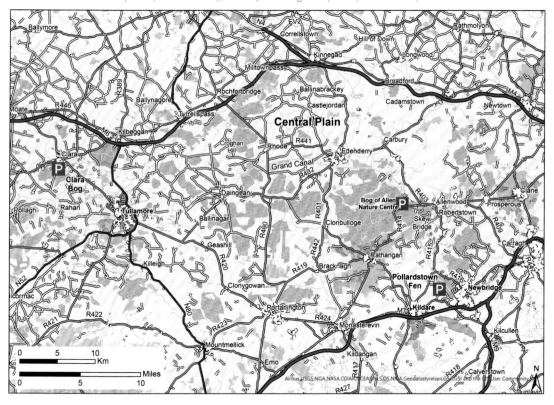

The Central Plain is a large, low-lying region formed by glacial drift underlain by limestone rocks, spanning Ireland from Galway to Dublin and dominated by the Shannon basin. The region is dominated by numerous raised bogs. The examples included here are those with good visitor facilities but there are several other interesting sites worthy of a visit in every county. A good starting place for information on all of Ireland's peatlands is the Irish Peatland Conservation Council's Bog of Allen Nature Centre in Lullymore, near Rathangan in County Kildare. Pollardstown Fen by the town of Kildare sixteen kilometres to the south is the largest remaining calcareous fen in Ireland. Further north in County Offaly is Clara Bog, a large raised bog with an award-winning visitor centre. To the south

Boardwalk, Abbeyleix Bog

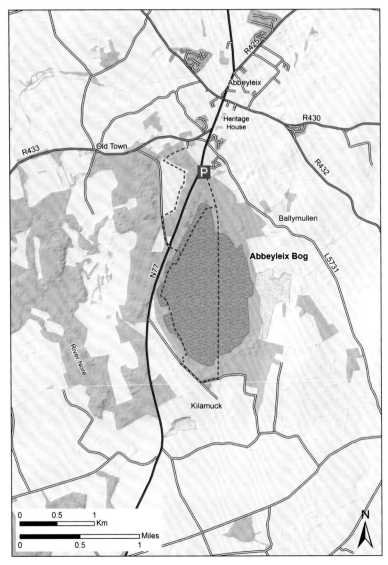

of Portlaoise is a formerly degraded, but well in recovery, raised bog at Abbeyleix in County Laois.

After the last Ice Age the retreating glaciers left many shallow lakes that initially filled in with fen and scrub. Later, sphagnum peat grew from the centres, creating large domed surfaces up to twelve metres thick. These accessible peatlands became the main economic resource for the region, supplying fuel for power stations and exported peat for horticultural use. The majority of the raised bogs were drained and stripped of their vegetation to extract the peat, but there are many scattered small remnants still clinging onto their special peatland wildlife. With the end of the commercial state-sponsored peat extraction, a new phase of peatland restoration has begun. One of the most ambitious initiatives is the Living Bog Partnership supported by the EU Life Programme and the Irish Government. Between 2016 and 2020, restoration works were carried out on twelve raised bog peatland sites across the central plains covering an area of 2,600 hectares.

The Bog of Allen is a complex of raised bogs that once covered 115,000 hectares across nine counties from the River Liffey in the east to the River Shannon in the west. At its centre lies the settlement of Lullymore, situated on an island of mineral soil. Part of the area called Lullymore Bog was one of the first commercially developed sites at the start of the mechanised industry, in the 20th century, supplying peat briquettes to fuel the nearby Allenwood power station. A small thirty-five-hectare section to the west of the nature centre, known as Lodge Bog, has been protected as a nature reserve and has a boardwalk across part of it.

Pollardstown Fen is a calcareous spring-fed fen formed in a post-glacial lake basin that remained too wet and alkaline to allow the normal transition to sphagnum bog. The fen supports a rich array of orchids, sedges and reeds. Agricultural drainage schemes in the 1960s damaged part of the site and two decades later most of the fen was purchased by the government's National Parks and Wildlife Service to help conserve it.

In 2000, plans for the construction of the Kildare bypass caused concern that the road would damage the aquifer that fed the fen, which is an internationally protected site. This famously led to the Taoiseach (prime minister) of the time mocking that a snail could hold up the building of a road. The snail in question was Geyer's whorl snail (*Vertigo geyeri*), a key indicator species for the health of the fens water table. The

Former commercial peat cutting site, Bog of Allen

road was designed to sit below the water table and sealed to allow the aquifer to rise to its normal levels, but snail numbers declined due to the temporary changes. Conservation bodies are working hard to help the snail population recover, including grazing parts of the site with Highland cattle.

Clara Bog, between the village of Ballycumber and the town of Clara in County Offaly, is largely state owned and is managed as a nature reserve. What was once a single large raised bog dome was bisected in the late 19th century by the construction of the Clara to Rahan Road. Associated drainage works resulted in the bog subsiding six metres and it now forms two separate, lower domes. In the 1980s the east part of the bog was drained as preparation for commercial peat cutting but appeals by conservationists including David Bellamy resulted in a reprieve and the development was stopped. The drains were later blocked and further restoration works carried out as part of the recent Living Bogs project. A one-kilometre boardwalk forms a loop round part of the reserve.

Abbeyleix Bog, sometimes referred to as Killamuck Bog, is a small raised bog that has been granted on a long-term lease to the local community by Bord na Móna. As with many other peatlands, the site was bisected by the construction of a railway in 1865. Being considered less valuable than the surrounding land made peatlands the target for the routing of transport infrastructure, despite the engineering difficulties of traversing deep peat. Parts of the bog edge were cut for domestic fuel and the estate owners planted large areas of woodland, including Scots pine in the 19th century. There are ancient oak woodlands in the vicinity, and around the wetter parts of the bog edges there are remnants of the original lagg fen scrub.

In the 1980s the site was acquired for commercial peat cutting and drainage works began but were halted after legal challenges from local residents. Recent management is aimed at ensuring the water table remains high and removing invading rhododendron and conifer seedlings from the recovering bog surface. There is a boardwalk on part of the site and the disused railway line provides further access.

The Abbeyleix Heritage Centre houses a museum explaining some of the area's peatland heritage, including a preserved cask of bog butter unearthed by peat workers from nearby Cashel Bog. That site also contained Europe's oldest fleshed bog body, dated as being a 4,000-year-old Bronze Age individual who appears to have been ritually killed. The body is now in the National Museum of Ireland in Dublin.

As well as the raised bogs, the region supports some good examples of blanket bog in Ireland's largest state-owned National Nature Reserve, Slieve Bloom Mountains, which rise from the flat plains between County Offaly and County Laois. Close to the

Drawing: Marsh fritillary

geographical centre of Ireland, these are among the oldest mountain ranges in Europe along with the Massif Central in France. Further west, next to the River Shannon, the Clonmacnoise monastic site founded in the 6th century is well worth a visit, with its views over the reedbeds of the Shannon Callows. The fen vegetation here originated from old peat workings and agricultural reclamation on previous raised bogs.

Travel Notes

Bog of Allen Visitor Centre can be reached by taking a bus from Dublin's Buáras to Allenwood and asking the driver to stop at the Skew Bridge over the canal. From here it is a four-kilometre walk to the centre on the right-hand side of the road at a junction.

Pollardstown Fen is three kilometres from Newbridge, which has a railway station. From Newbridge take the L7037 Standhouse Road to the Curragh racecourse. Take the second turning on the right after the Cill Mhuire Church, onto the L7032. The reserve car park is 1.5 kilometres on the right and from there a trail and boardwalk lead onto the south-west section of the fens.

Abbeyleix Bog is located one kilometre south along the Old Dublin to Cork Road from Abbeyleix Heritage centre. The Manor Hotel is the last building on the left as you leave Abbeyleix town on the N8 heading towards Cork. Leave the hotel car park through the gap in the beech hedge, turn left to reach an interpretative board marking the entrance to the bog. There is also a sign for Killamuck Bog.

Clara Bog Visitor Centre is in the library in the town of Clara, which has a railway station. From the centre the bog is two kilometres south on the Clara to Rahan Road with a small car park leading to a looped boardwalk trail.

Slieve Bloom Mountains can be accessed from a large car park at N 364 065, with trails and a viewing platform at the Ridge of Capard, signposted from Rosenallis village twenty kilometres north-west of Portlaoise. The Slieve Bloom Way is a seventy-five-kilometre long-distance trail around the mountain range that passes through the nature reserve.

Place names

Lullymore: from Irish *Loilgheach Mór*, meaning 'big place for cows to calve'.

Abbeyleix: from nearby Abbey of Leix, founded in the year 600 CE.

Clara: from Irish *claragh*, meaning 'a level place'.

Killamuck: from Irish *Coill na Muc*, meaning 'wood of the swine'.

Slieve Bloom: from Irish *Sliabh Bladhma*, meaning 'the mountain of *Bladh*' (an ancient Connaught folk hero).

WICKLOW MOUNTAINS

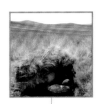

Map: OSI 1:50,000 Sheet Nos 56, 62
Peatland Grid ref: Glenealo Valley T 072 965, Liffey Head Bog O 137 127
Access point: Glendalough Visitor Centre car park T 124 968, Loch Tay car park for Djouce Mountain (O 161 081)

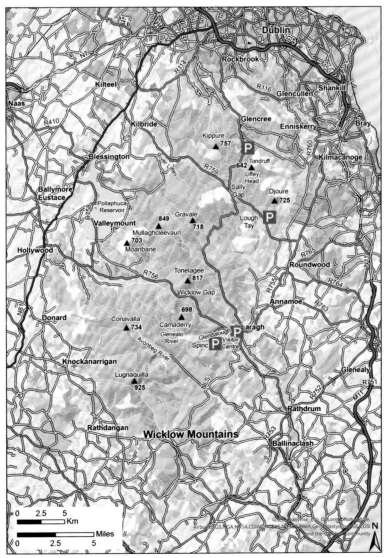

The wild expanse of the Wicklow Mountains, 'the Garden of Ireland', lies a short distance south of Dublin and forms the largest continuous upland area in Ireland. Situated in the

Peat pipe erosion feature, Wicklow Mountains

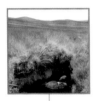

Ruin on St Kevin's Way, Wicklow Gap

drier climate of the east, the gently undulating granite-formed mountain summits create their own wet conditions that favour blanket bog. Some of the best areas are found in a hollow between Djouce, Kippure and Tonduff mountain at the head of the Liffey River and in the upper reaches of Glendalough where the main visitor centre of the Wicklow National Park is located.

The name Wicklow probably comes from the Old Norse word *Vykyngelo*, meaning 'meadow of the Vikings'. The Irish have an alternative name *Cill Mhantáin*, meaning 'church of the toothless one', from the story of Saint Patrick attempting to land his ship in the area and being attacked by locals, resulting in one of his party losing a tooth in the fighting. When he returned later to set up a church there, he named it *Manntach*, 'the toothless'.

The peatlands here have long been cut for fuel but the construction of the Great Military Road from 1800 to 1809, to neutralise the mountain wilderness as a refuge for those hiding out after the 1798 Rebellion, also gave local people greater access to the peatlands and extended the amount of turf cutting. During the Second World War, peat cutting expanded to provide fuel for the people of Dublin. It is humbling to think of the city dwellers making the long journey up the steep hills to spend days hand cutting the peat and letting it dry before bringing it home for their fires.

In the mid-1980s mechanised cutting and drainage were threatening large areas of blanket bog. At Liffey Head Bog, drainage in preparation for peat cutting led to protests by conservationists. The government acquired the area, and it became a core area for the formation of the Wicklow National Park. In 1995 work began on blocking the drains to restore the peatland. The military road running north-south through this area between Glencree and the famous viewpoint at Sally Gap, offers a captivating panorama across the isolated peatlands. The peatlands can also be viewed from the summit of Djouce mountain, reached from the south by a trail that includes a long boardwalk made of railway sleepers to protect the soft, wet ground. Alongside blanket bog the Liffey Head Bog contains mineral-rich flushes that support plants more typical of fens, such as star sedge (*Carex echinata*) and rushes. Wicklow Mountains is also one of the best areas in Ireland to see red grouse.

Further south, in Glendalough, much of the surrounding blanket bog was planted with commercial conifer forest. Some of the most important remaining areas are in the Glenealo Valley National Nature Reserve. There is a challenging ten-kilometre walking trail, Spinc and Glenealo Valley Loop, with steep sections of wooden boardwalk – well worth the climb for the views. Part of the route leads down through blanket bog and past the ruins of a lead mining village that operated until 1925.

Travel Notes

The city of Dublin has a main railway station and there are buses from St. Stephen's Green to the visitor centre at Glendalough with waymarked trails to Glenealo valley.

The Wicklow Way is a 132-kilometre long-distance walking route that leads from Dublin down the length of the Wicklow Mountains using minor roads and moorland tracks. There is access to the east side of Djouce Mountain for a boardwalk trail up to the summit and views across Liffey Head Bog. The alternative is to follow the Great Military Road which makes for a challenging but rewarding cycle to Sally Gap from the hamlet of Glencree.

Place names

Tonduff: from Irish *toin dubh*, meaning 'black backside'.

Glendalough: from Irish *Gleann Dá Loch*, 'valley of two lakes'.

Glenealo: from Irish Glanaslagh, meaning 'the valley of the waterfall of the hill'.

Spinc: from the Irish *An Spinc*, meaning 'pointed hill'.

Djouce: from Irish *Dioghais*, meaning 'fortified height'.

Sally Gap: from Irish *saile*, meaning 'willow'.

Safety and access

Scottish Outdoor Access code

https://www.outdooraccess-scotland.scot/

Safety

Peatlands are a potentially hazardous environment for recreational users. Navigation can be difficult over unfamiliar peatlands as they have few reference features, and it is easy to get disorientated or lost. Deep open water, ditches and saturated soils can be dangerous, but the risks can be overcome by sticking to marked routes. Most of the sites in this guide are in remote and often exposed mountainous or hilly areas susceptible to severe weather, especially in winter. Visits need to be properly planned and visitors should have the right clothing and equipment. Advice on mountain and hill walking safety is provided by The Ramblers.

Peatland vegetation and soils are fragile and easily damaged by trampling. Most of the peatlands have good access facilities with trails on stable ground or with access assisted by boardwalks. It is important to remain on these routes and to keep dogs in close control or on leads.

The adder is the UK's only venomous snake and is found on peat bogs across Britain but is absent from Ireland. Adders are timid reptiles and unlikely to bite unless threatened. Most bites happen when people get too close or try to handle the snakes. If found in the wild they should be left alone and will move away of their own accord. Further advice on adders and what to do in cases where a person or pet gets bitten is available from the Amphibian and Reptile Conservation Trust.

English Outdoor Access code

www.gov.uk/countryside-code

SELECTED BIBLIOGRAPHY

Bonn, A. et al (eds) 2016. Peatland Restoration and Ecosystem Services: Science, Policy, and Practice. Cambridge University Press, Cambridge.

Buckland, P. (ed) 2014. Thorne and Hatfield Moors Papers Volume 9. Thorne and Hatfield Moors Conservation Forum, Doncaster.

Caufield, C. 1991. Thorne Moors. Sumach Press, St. Albans.

Charman, D. 2002. Peatlands and Environmental Change. Wiley, Chichester.

Darby, H.C. 1940. The Medieval Fenland. Cambridge University Press, Cambridge.

Darby, H.C. 1956. The Draining of the Fens. 2nd ed. Cambridge University Press, Cambridge

Glob, P.V. 1971. The Bog People: Iron-Age Man Preserved. Paladin, London.

Håkan, R., Jeglum, J.K. 2013. The Biology of Peatlands. 2nd ed. Oxford University Press, Oxford.

Lindsay, R. 1995. Bogs: The Ecology, Classification and Conservation of Ombrotrophic Mires. Scottish Natural Heritage, Perth.

Lindsay, R. 2016. 'Peatland (Mire Types): Based on Origin and Behavior of Water, Peat Genesis, Landscape Position, and Climate.' in: Finlayson, C. Max, Milton, G. Randy, Prentice, R. Crawford and Davidson, Nick C. (ed.) The Wetland Book: II: Distribution, Description and Conservation. Springer, Netherlands.

Lindsay, R. and Andersen, R. 2016. 'Blanket Mires of Caithness and Sutherland, Scotland's Great Flow Country (UK).' in: Finlayson, C. Max, Milton, G. Randy, Prentice, R. Crawford and Davidson, Nick C. (ed.) The Wetland Book: II: Distribution, Description and Conservation. Springer Netherlands. pp. 1-17.

Lindsay, R., Charman, D.J., Everingham, F., O'Reilly, R.M., Palmer, M.A., Rowell, T.A. and Stroud, D.A. 1988. The Flow Country: The peatlands of Caithness and Sutherland. Edited by D.A. Ratcliffe and P. Oswald. Nature Conservancy Council, Peterborough.

Lindsay, R.A. and Clough, J. 2017. 'United Kingdom.' in: H. Joosten, F. Tanneberger and A. Moen (eds.) Mires and peatlands of Europe – Status, distribution and conservation. pp. 705-720. Schweizerbart Science Publishers, Stuttgart.

McBride, A., Diack, I., Droy, N., Hamill, B., Jones, P., Schutten, J., Skinner, A. & Street, M. (eds.) (2011) The Fen Management Handbook. p. 332. Scottish Natural Heritage, Perth.

Moore, P.D. and Bellamy, D.J. 1974. Peatlands. Elek Science, London.

O'Connell, C. (ed) 1987. The IPCC Guide to Irish Peatlands. Irish Peatland Conservation Council, Dublin.

O'Reilly, J., O'Reilly, C. & Tratt, R. 2012. Field Guide to Sphagnum Mosses in Bogs. Field Studies Council, Shrewsbury.

Stroud, D., Reed, T., Pienkowski M.W. & Lindsay, R. 1987. Birds, Bogs and Forestry. The Peatlands of Caithness and Sutherland. Nature Conservancy Council, Edinburgh.

Stroud, D., Pienkowski, M.W., Reed, T. and Lindsay, R. 2015. 'The Flow Country – battles fought, war won, organisation lost.', in: Thompson, D., Birks, H. and Birks, J. (eds.) Nature's Conscience – The life and legacy of Derek Ratcliffe. Langford Press, Peterborough pp. 401-439.

Thom, T., Hanlon, A., Lindsay, R., Richards, J., Stoneman, R. and Brooks, S. 2019. Conserving Bogs: The Management Handbook. IUCN UK Peatland Programme, Edinburgh.

Useful websites

IUCN UK Peatland Programme (www.iucn-uk-peatlandprogramme.org)

Peatland Projects Map (www.iucn-uk-peatlandprogramme.org/projects-map)

Brecon Beacons National Park (www.breconbeacons.org)

Butterfly Conservation (www.butterfly-conservation.org)

Cairngorms National Park (www.cairngorms.co.uk)

Connemara National Park (www.connemaranationalpark.ie)

Dartmoor National Park (www.dartmoor.gov.uk)

Department of Agriculture, Environment and Rural Affairs Northern Ireland (www.daera-ni.gov.uk)

Department for Environment Food and Rural Affairs (www.gov.uk/government/organisations/department-for-environment-food-rural-affairs)

Exmoor National Park (www.exmoor-nationalpark.gov.uk)

Humberhead Peatlands (www.humberheadpeatlands.org.uk)

Irish Peatland Conservation Council (www.ipcc.ie)

Loch Lomond and Trossachs National Park (www.lochlomond-trossachs.org)

Marble Arch Caves Geopark (www.marblearchcavesgeopark.com)

Marches Mosses BogLIFE (www.themeresandmosses.co.uk)

Moors for the Future Partnership (www.moorsforthefuture.org.uk)

National Museum of Ireland (www.museum.ie)

National Trust (www.nationaltrust.org.uk)

National Trust for Ireland (https://www.antaisce.org/)

National Trust for Scotland (www.nts.org.uk)

National Parks and Wildlife Service Ireland (www.npws.ie)

Natural England (www.gov.uk/government/organisations/natural-england)

Natural Resources Wales (www.naturalresourceswales.gov.uk)

NatureScot (www.nature.scot)

North Pennines AONB, Pennine PeatLIFE (www.northpennines.org.uk)

Peatland Action (www.nature.scot/climate-change/nature-based-solutions/peatland-action-project)

Plantlife (www.plantlife.org.uk)

Ramblers (www.ramblers.org.uk)

RSPB (www.rspb.org.uk)

Scottish Outdoor Access Code (www.outdooraccess-scotland.scot)

Scottish Wildlife Trust (www.scottishwildlifetrust.org.uk)

Solway Wetlands (www.solwaywetlands.org.uk)

The Great Fen Project (www.greatfen.org.uk)

The Peatlands Partnership (https://www.theflowcountry.org.uk/about-us/the-peatlands-partnership)

The Wildlife Trusts (www.wildlifetrusts.org)

Thorne and Hatfield Moors Conservation Forum (www.thmcf.org)

Ulster Wildlife Trust (www.ulsterwildlife.org)

Wild Nephin Ballycroy National Park (www.wildnephinnationalpark.ie)

Wicklow Mountains National Park (www.wicklowmountainsnationalpark.ie)

Yorkshire Peat Partnership (www.yppartnership.org.uk)

Every effort has been made by the author and publisher to confirm the information in this book is accurate and they accept no responsibility for any loss, injury or inconvenience experienced by any person or persons whilst using this book.

Photo credits

Drawings and paintings by Darren Rees page 74, page 92, page 112, page 132, page 139, page 146, page 171, page 190, page 208, page 218, page 220, page 246

Photography by the Author except for the following:

Laurie Cambell – Front cover – Peatbog Lochan, back cover – Glen Affric.

Karin Barancova – endpiece (the author)

Jason Bye – page 3 (Tony Juniper)

Brian Eversham – page 33 (four-spotted chaser), (mire pill-beetle), (bog bush-cricket), (large red damselfly), (common hawker), (large heath), (black darter female) (black darter male), page 38 (early marsh-orchid x southern marsh-orchid), page 175 (Holme Fen post)

Tina Claffey – page 34 (raft spider)

Norman Russel – page 35 (golden plover), page 62 (Sphagnum and felled trees)

National Museum of Ireland – page 41 (Clonycavan man)

Moors for the Future Partnership – page 65 (Black Hill pre-restoration and post-restoration), page 136 (Black Hill Trig Point, Pennine Way)

Penny Anderson – page 64 (wooden dam, Kinder Scout)

Christine Hall/RSPB – page 77 (View from summit of Trumland Hill, Rousay looking East, north-east to Egilsay and Eday), page 84 (View up the Post Road glen towards the Cuilags).

Blue Kirkhope – page 78 (Hermaness National Nature Reserve), page 81 (Hermaness NNR)

Lorne Gill – page 82 (Hermaness), page 110 (Rannoch Moor with Corrour Station)

RSPB – page 86 (Wilderness Track and peat cutting Birsay Moors, Orkney), page 87 (Eddie Balfour Hide, Cottascarth, Orkney), page 167 (Bowness Common), page 172 (Bowness Common)

Benjamin Inglis-Grant – page 97 (looking south to Harris near Lochganvich, Lewis), page 98 (eroding peat hags, Bheinn Bhragair , Lewis), page 102 (peat stacks on Pentland Road, Lewis)

Dom Hinchley – page 142 (Grinton Moor looking across to Reeth)

Lyndon Marquis – page 145 (Greensett Moss from the flank of Whernside looking across Chapel le Dale to Ingleborough)

Robert Duff/Natural England – page 160 (Fenn's Moss), page 163 (Fenn's Moss), page 164 (bunded area of Bettisfield Moss)

RSPB-Images – page 168 (Bowness Common)

Natural England – page 170 (Glasson Moss observation tower)

The Wildlife Trust for Bedfordshire, Cambridgeshire & Northamptonshire - page 180 (Woodwalton Fen)

Mike Perks/RSPB – page 193 (Hafod, Lake Vyrnwy), page 200 (top of Rhiwargor Falls, Lake Vyrnwy)

Pete Jones/NRW – page 194 (fly orchid, Cors Bodeilio, page 196 (Cors Goch), page 197 (fen meadow, Cors Bodeilio), page 198 (Cors Bodeilio)

Gethin Elias/RSPB – page 202 (Bryn Mawr, Lake Vyrnwy)

Natural Resources Wales – page 215 (Ogof Ffynnon Du, Brecon Beacons), page 216 (Pant-y-Sais), page 216 (Cors Crymlyn)

Stuart Housden – back flap